McGraw-Hill
Mathematics

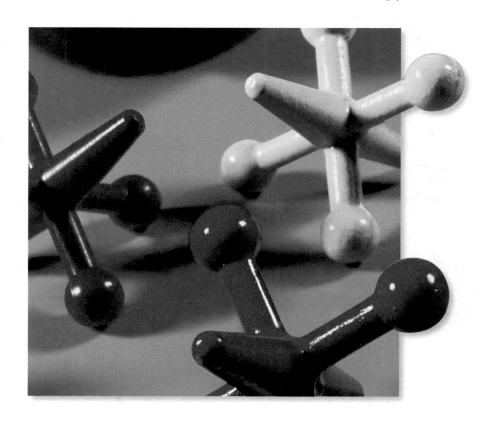

**McGraw-Hill
School Division**

New York   Farmington

Senior Program Authors

Gunnar Carlsson, Ph.D.
Professor of Mathematics
Stanford University
Stanford, California

Ralph L. Cohen, Ph.D.
Professor of Mathematics
Stanford University
Stanford, California

Program Authors

Douglas H. Clements, Ph.D.
Professor of Mathematics Education
State University of New York at Buffalo
Buffalo, New York

Lois Gordon Moseley, M.S.
Mathematics Consultant
Houston, Texas

Robyn R. Silbey, M.S.
Montgomery County Public Schools
Rockville, Maryland

Carol E. Malloy, Ph.D.
Assistant Professor of Mathematics Education
University of North Carolina at Chapel Hill
Chapel Hill, North Carolina

McGraw-Hill School Division
A Division of The McGraw-Hill Companies

Copyright © 2002 McGraw-Hill School Division,
a Division of the Educational and Professional Group of The McGraw-Hill Companies, Inc.
All Rights Reserved.

McGraw-Hill School Division
Two Penn Plaza
New York, New York 10121-2298

Printed in the United States of America
ISBN 0-02-100125-1
1 2 3 4 5 6 7 8 9 073/043 05 04 03 02 01 00

Contributing Authors

Mary Behr Altieri, M.S.
Mathematics Teacher
1993 Presidential Awardee
Lakeland Central School District
Shrub Oak, New York

Nadine Bezuk, Ph.D.
Professor of Mathematics Education
San Diego State University
San Diego, California

Pam B. Cole, Ph.D.
Associate Professor of
Middle Grades English Education
Kennesaw State University
Kennesaw, Georgia

Barbara W. Ferguson, Ph.D.
Assistant Professor of Mathematics
and Mathematics Education
Kennesaw State University
Kennesaw, Georgia

Carol P. Harrell, Ph.D.
Professor of English and English Education
Kennesaw State University
Kennesaw, Georgia

Donna Harrell Lubcker, M.S.
Assistant Professor of Education
and Early Childhood
East Texas Baptist University
Marshall, Texas

Chung-Hsing OuYang, Ph.D.
Assistant Professor of Mathematics
California State University, Hayward
Hayward, California

Marianne Weber, M.ED.
National Mathematics Consultant
St. Louis, Missouri

Contents

Chapter 1: Addition and Subtraction Strategies and Facts to 12

Theme: At the Pond

Chapter 2:
Addition and Subtraction Strategies and Facts to 20

Theme: Postcards from Pete

Chapter 3: Place Value

Theme: Snap and Tap!

Chapter 4: Money

Theme: At the Fair

Chapter 5:
Add 2-Digit Numbers
Theme: People Together

Chapter 6:
Subtract 2-Digit Numbers
Theme: Seasons

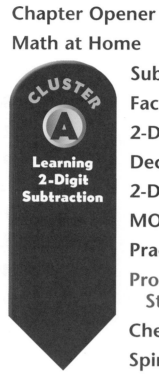

Chapter 7:
Time

Theme: It's Show Time!

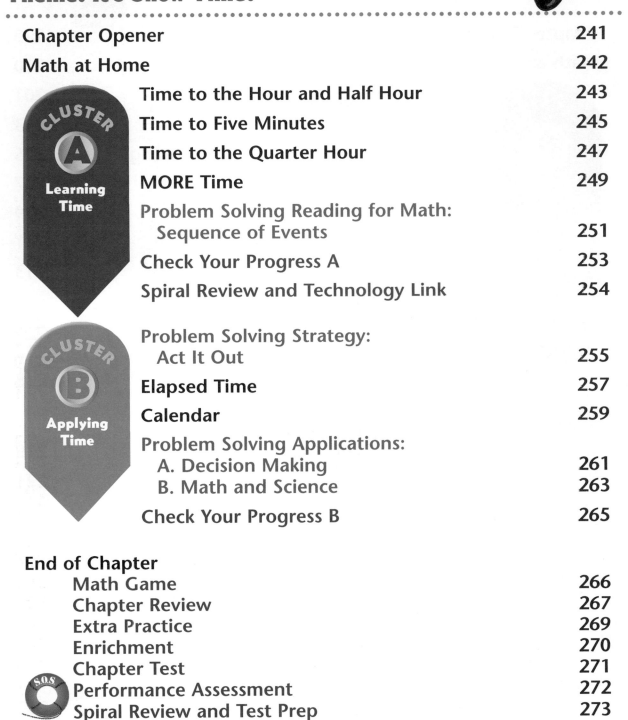

Chapter 8:
Data and Graphs
Theme: All About Us

CLUSTER **A** Graphs

CLUSTER **B** Applying Data and Graphs

Chapter 9: Measurement

Theme: Long, Long Ago

Chapter 10: Geometry

Theme: Cityscapes

Chapter 11:
Fractions and Probability
Theme: Celebrate the Sea!

Chapter 12:
Place Value to 1,000

Theme: Beads, Buttons, and Things

Chapter 13:
Add and Subtract 3-Digit Numbers

Theme: Community Helpers

Chapter 14:
Multiplication and Division
Theme: Coast to Coast

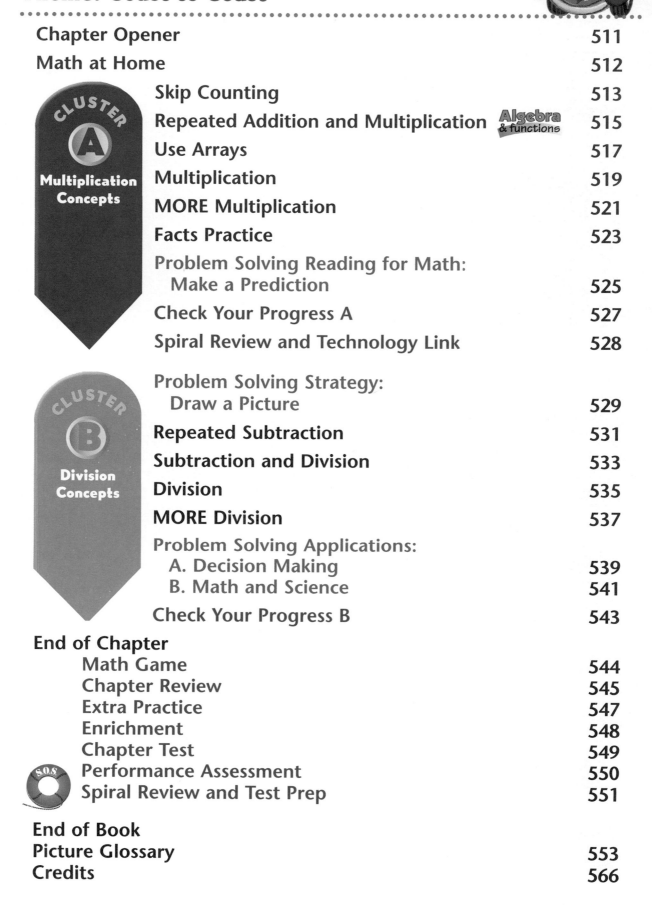

CHAPTER 1

Addition and Subtraction Strategies and Facts to 12

theme
At the Pond

Use the Data

How many different kinds of animals can you find?

Tell an addition story.

Tell a subtraction story.

What You Will Learn

In this chapter you will learn how to:

- Add, facts to 12.

- Subtract, facts to 12.

- Draw pictures to solve problems.

By the Pond

Story by Becky Manfredini
Illustrated by Bernard Adnet

McGraw-Hill
School Division

Math Words

sum

Add to find the sum.
3 + 2 = 5 ◄——— sum

turnaround facts

$$\begin{array}{r} 4 \\ +3 \\ \hline 7 \end{array} \qquad \begin{array}{r} 3 \\ +4 \\ \hline 7 \end{array}$$

difference

6 - 3 = 3 ◄——— difference

fact family

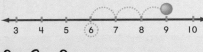

5 + 1 = 6	6 - 5 = 1
1 + 5 = 6	6 - 1 = 5

count on

Start with 6 and count on 3.

3 4 5 6 7 8 9 10

6 + 3 = 9

Dear Family,

In Chapter 1, I will learn strategies for adding and subtracting numbers to 12. Here are new vocabulary words and an activity that we can do together.

Button Addition

- Place some buttons on the table. Have your child count how many.

- Place another button on the table. Ask how many there are now.

  For example, 7 + 1 = 8.

- Place two more buttons on the table and have your child find how many in all.

  For example, 8 + 2 = 10.

use

12 buttons or pennies

Additional activities at
www.mhschool.com/math

Name _____

Learn

Start with the greater number.

You can use a number line to count on to add.

Add 8 + 3.

Math Words

number line
count on
add

0 1 2 3 4 5 6 7 8 9 10 11 12

8 + 3 = __11__

Start at 8.   Count on 3.

Try it   Add. You can use the number line.

0 1 2 3 4 5 6 7 8 9 10 11 12

1. 3 + 1 = __4__   5 + 2 = ___   12 + 0 = ___

2. 8 + 2 = ___   7 + 1 = ___   4 + 3 = ___

3.
8   5   4   10   6   3   9
+3   +1   +0   +1   +2   +2   +3

4.
6   7   2   6   9   1   3
+0   +2   +9   +3   +2   +2   +7

Sum it Up   How can using a number line help you to add 7 + 3?

Math at Home: Your child used a number line to add on 0, 1, 2, or 3 to a number.
Activity: Say an addition fact that adds on 1, 2, or 3, such as 5 + 2. Have your child count on to find the sum.

three   3

Practice Add.

Start with the greater number.

5. 6 + 2 = _8_   7 + 0 = ___   5 + 3 = ___

6. 4 + 1 = ___   6 + 3 = ___   8 + 1 = ___

7. 3 + 5 = ___   9 + 2 = ___   7 + 5 = ___

8. 5 + 2 = ___   9 + 3 = ___   3 + 0 = ___

9.
$$\begin{array}{c} 9 \\ +\,1 \\ \hline \end{array} \qquad \begin{array}{c} 9 \\ +\,2 \\ \hline \end{array} \qquad \begin{array}{c} 10 \\ +\,0 \\ \hline \end{array} \qquad \begin{array}{c} 7 \\ +\,3 \\ \hline \end{array} \qquad \begin{array}{c} 8 \\ +\,2 \\ \hline \end{array} \qquad \begin{array}{c} 5 \\ +\,0 \\ \hline \end{array} \qquad \begin{array}{c} 3 \\ +\,5 \\ \hline \end{array}$$

10.
$$\begin{array}{c} 0 \\ +\,8 \\ \hline \end{array} \qquad \begin{array}{c} 7 \\ +\,3 \\ \hline \end{array} \qquad \begin{array}{c} 10 \\ +\,2 \\ \hline \end{array} \qquad \begin{array}{c} 8 \\ +\,2 \\ \hline \end{array} \qquad \begin{array}{c} 4 \\ +\,2 \\ \hline \end{array} \qquad \begin{array}{c} 11 \\ +\,0 \\ \hline \end{array} \qquad \begin{array}{c} 9 \\ +\,3 \\ \hline \end{array}$$

Problem Solving

Mental Math

11. The frog takes 7 hops. Then it takes 2 more hops. How many hops does the frog take in all?

_____ hops

12. There are 5 green bugs on the log. One more bug lands on the log. How many bugs are on the log now?

_____ bugs

Name _____

Learn

The sum does not change when you turnaround the addends.

Math Words

turnaround facts
addend
sum

These are turnaround facts.

$4 + 7 = 11$

$7 + 4 = 11$

Try it   Write two facts for each picture.

1. ___1___ + ___3___ = ___4___   ___3___ + ___1___ = ___4___

2. ___ + ___ = ___   ___ + ___ = ___

3. ___ + ___ = ___   ___ + ___ = ___

How are $3 + 5 = 8$ and $5 + 3 = 8$ alike? How are they different?

Math at Home: Your child learned that the sum does not change when you turn around addends, such as $3 + 5 = 8$ and $5 + 3 = 8$.
Activity: Ask your child to use buttons to show the turnaround facts for the addends 4 and 2.

McGraw-Hill School Division

five **5**

4.

| 3
+ 6 | 6
+ 3 | 1
+ 5 | 5
+ 1 | 4
+ 5 | 5
+ 4 |

9   *9*

5.

| 2
+ 8 | 8
+ 2 | 9
+ 3 | 3
+ 9 | 5
+ 6 | 6
+ 5 |

6.

| 5
+ 7 | 7
+ 5 | 6
+ 4 | 4
+ 6 | 2
+ 9 | 9
+ 2 |

Are these turnaround facts? Why or why not?
Write yes or no.

7.

6 + 5   5 + 7   8 + 3

5 + 6   5 + 5   3 + 8

_____   _____   _____

8. Use the numbers 3, 7, and 4.
Write two addition facts that are
turnaround facts.

_____

Name _____

Add.

1. $3 + 1 = $ _____

2. $3 + 4 = $ _____

3. $9 + 1 = $ _____

4. $2 + 2 = $ _____

5. $3 + 7 = $ _____

6. $3 + 8 = $ _____

7. $4 + 8 = $ _____

8. $9 + 2 = $ _____

9. $7 + 4 = $ _____

10. $9 + 0 = $ _____

11. $4 + 4 = $ _____

12. $4 + 3 = $ _____

13. $3 + 2 = $ _____

14. $3 + 6 = $ _____

15. $5 + 6 = $ _____

16. $1 + 3 = $ _____

17. $4 + 6 = $ _____

18. $6 + 3 = $ _____

19. $6 + 2 = $ _____

20. $0 + 5 = $ _____

21. $7 + 3 = $ _____

22. $2 + 4 = $ _____

23. $8 + 0 = $ _____

24. $2 + 8 = $ _____

Math at Home: Your child practiced addition facts to 12.
Activity: Cover the answers with a paper strip. Time your child as he or she writes the answers again. You can repeat daily to help your child recall the facts quickly.

Facts Practice: Addition

Add.

1.
$$\begin{array}{r} 2 \\ +7 \\ \hline \end{array} \quad \begin{array}{r} 5 \\ +5 \\ \hline \end{array} \quad \begin{array}{r} 6 \\ +2 \\ \hline \end{array} \quad \begin{array}{r} 3 \\ +5 \\ \hline \end{array} \quad \begin{array}{r} 3 \\ +1 \\ \hline \end{array} \quad \begin{array}{r} 6 \\ +1 \\ \hline \end{array} \quad \begin{array}{r} 8 \\ +3 \\ \hline \end{array}$$

2.
$$\begin{array}{r} 7 \\ +2 \\ \hline \end{array} \quad \begin{array}{r} 9 \\ +1 \\ \hline \end{array} \quad \begin{array}{r} 4 \\ +1 \\ \hline \end{array} \quad \begin{array}{r} 2 \\ +2 \\ \hline \end{array} \quad \begin{array}{r} 5 \\ +0 \\ \hline \end{array} \quad \begin{array}{r} 8 \\ +2 \\ \hline \end{array} \quad \begin{array}{r} 3 \\ +9 \\ \hline \end{array}$$

3.
$$\begin{array}{r} 4 \\ +4 \\ \hline \end{array} \quad \begin{array}{r} 8 \\ +4 \\ \hline \end{array} \quad \begin{array}{r} 2 \\ +3 \\ \hline \end{array} \quad \begin{array}{r} 1 \\ +2 \\ \hline \end{array} \quad \begin{array}{r} 6 \\ +3 \\ \hline \end{array} \quad \begin{array}{r} 3 \\ +3 \\ \hline \end{array} \quad \begin{array}{r} 2 \\ +7 \\ \hline \end{array}$$

4.
$$\begin{array}{r} 1 \\ +1 \\ \hline \end{array} \quad \begin{array}{r} 2 \\ +1 \\ \hline \end{array} \quad \begin{array}{r} 4 \\ +3 \\ \hline \end{array} \quad \begin{array}{r} 5 \\ +2 \\ \hline \end{array} \quad \begin{array}{r} 7 \\ +1 \\ \hline \end{array} \quad \begin{array}{r} 8 \\ +3 \\ \hline \end{array} \quad \begin{array}{r} 7 \\ +4 \\ \hline \end{array}$$

5.
$$\begin{array}{r} 1 \\ +3 \\ \hline \end{array} \quad \begin{array}{r} 8 \\ +1 \\ \hline \end{array} \quad \begin{array}{r} 9 \\ +2 \\ \hline \end{array} \quad \begin{array}{r} 9 \\ +3 \\ \hline \end{array} \quad \begin{array}{r} 5 \\ +4 \\ \hline \end{array} \quad \begin{array}{r} 0 \\ +2 \\ \hline \end{array} \quad \begin{array}{r} 2 \\ +6 \\ \hline \end{array}$$

6.
$$\begin{array}{r} 3 \\ +2 \\ \hline \end{array} \quad \begin{array}{r} 7 \\ +3 \\ \hline \end{array} \quad \begin{array}{r} 6 \\ +5 \\ \hline \end{array} \quad \begin{array}{r} 6 \\ +6 \\ \hline \end{array} \quad \begin{array}{r} 3 \\ +6 \\ \hline \end{array} \quad \begin{array}{r} 5 \\ +7 \\ \hline \end{array} \quad \begin{array}{r} 1 \\ +8 \\ \hline \end{array}$$

7.
$$\begin{array}{r} 2 \\ +6 \\ \hline \end{array} \quad \begin{array}{r} 1 \\ +7 \\ \hline \end{array} \quad \begin{array}{r} 3 \\ +4 \\ \hline \end{array} \quad \begin{array}{r} 4 \\ +8 \\ \hline \end{array} \quad \begin{array}{r} 6 \\ +4 \\ \hline \end{array} \quad \begin{array}{r} 3 \\ +7 \\ \hline \end{array} \quad \begin{array}{r} 2 \\ +9 \\ \hline \end{array}$$

Name _____

Read to Set a Purpose

Reading Skill   You can look for information to help you solve a story problem.

Turtles like sunny days
at the pond.
4 turtles are on the log.
5 turtles are swimming.

Solve.

1. How many turtles are on the log? _____ turtles

2. How many turtles are swimming? _____ turtles

3. How many turtles are at the pond? _____ turtles

Write a number sentence to show your thinking.

_____

4. What information helped you find the answer?

_____

 Math at Home: Your child read a story and set a purpose for reading.
Activity: Tell your child a simple number story. Then have your child identify the purpose of the numbers used in the story.

McGraw-Hill School Division

Solve.

Laura saw 3 bluebirds
near the pond.
Ling saw 4 yellow birds
in a tree.

5. How many blue birds did Laura see near the pond?   _____

6. How many yellow birds did Ling see in the tree?   _____

7. How many birds did they see altogether?   _____ birds

Write a number sentence to show your thinking.

_____

8. What information helped you to find the answer?

_____

9. Use the story. Write a problem.

_____

Name _____

Add. You can use the number line.

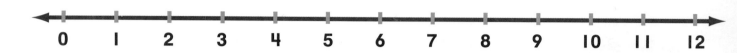

0 1 2 3 4 5 6 7 8 9 10 11 12

1. $0 + 4 =$ _____ $3 + 9 =$ _____ $1 + 3 =$ _____

2. $6 + 2 =$ _____ $9 + 1 =$ _____ $5 + 3 =$ _____

3.
$$\begin{array}{r} 7 \\ +2 \\ \hline \end{array} \qquad \begin{array}{r} 3 \\ +1 \\ \hline \end{array} \qquad \begin{array}{r} 5 \\ +3 \\ \hline \end{array} \qquad \begin{array}{r} 9 \\ +2 \\ \hline \end{array} \qquad \begin{array}{r} 7 \\ +3 \\ \hline \end{array} \qquad \begin{array}{r} 8 \\ +2 \\ \hline \end{array}$$

Add.

4.
$$\begin{array}{r} 2 \\ +2 \\ \hline \end{array} \qquad \begin{array}{r} 2 \\ +3 \\ \hline \end{array} \qquad \begin{array}{r} 5 \\ +5 \\ \hline \end{array} \qquad \begin{array}{r} 5 \\ +6 \\ \hline \end{array} \qquad \begin{array}{r} 3 \\ +3 \\ \hline \end{array} \qquad \begin{array}{r} 3 \\ +4 \\ \hline \end{array}$$

5.
$$\begin{array}{r} 4 \\ +4 \\ \hline \end{array} \qquad \begin{array}{r} 4 \\ +5 \\ \hline \end{array} \qquad \begin{array}{r} 1 \\ +6 \\ \hline \end{array} \qquad \begin{array}{r} 6 \\ +1 \\ \hline \end{array} \qquad \begin{array}{r} 8 \\ +4 \\ \hline \end{array} \qquad \begin{array}{r} 4 \\ +8 \\ \hline \end{array}$$

Journal

6. Write two turnaround addition facts for the cube train.

_____ + _____ = _____

_____ + _____ = _____

Count how many.

1.  _____

_____

Look at each pattern. What could the next shape be?

2.  _____

_____

Write each number.

3. fifteen _____ twenty-two _____

TECHNOLOGY LINK

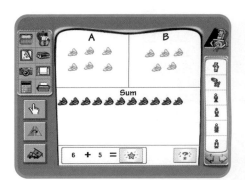

Model Addition
- Use counters.
- Choose a mat to add.
- Stamp out 6 butterflies.
- Stamp out 5 butterflies.
- Click on +.
- What fact is shown? _____
  1. Use counters. Add 7 and 4.
     What fact is shown? _____
  2. Stamp out other addition facts.

For more practice use Math Traveler™.

Name _____

What do I know?

What do I need to find out?

Draw a Picture

Read ▶ There are 7  at the pond.

3 more 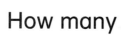 come to the pond.

How many are at the pond now?

Plan ▶ I can draw a picture to solve.

Solve ▶

Look Back ▶ Does my answer make sense? Why?

Draw a picture. Solve.

1. There are 3 making a dam.

Another  joins them.

How many are there in all?

_____

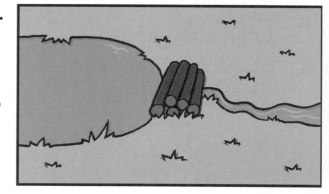

 How did your picture help you solve the problem?

Math at Home: Your child solved problems by drawing a picture.
Activity: Tell a simple addition problem for your child to solve. For example,
"I found 5 pennies. You found 6 pennies. How many pennies did we find?"

Draw a picture. Solve.

2. Ned counts 8 birds.
Then he counts 2 more.
How many birds does
he count in all?

_____ birds

3. There are 3 turtles
on the log. 3 more turtles
climb onto the log.
How many turtles are on
the log?

_____ turtles

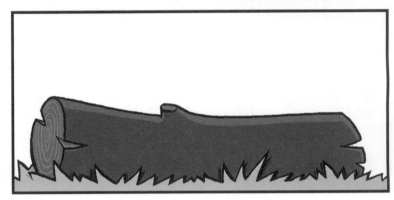

4. Mandy catches 4 fish.
Her brother catches the
same number of fish.
How many fish do they
catch altogether?

_____ fish

Critical Thinking   Journal

5. Look at the picture.
Write an addition problem.

_____

Name _____

Learn

You can use a number line to count back to subtract. Count back to find the difference.

Math Words

count back
subtract
difference

Find 12 − 3.

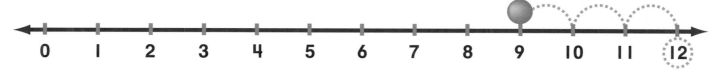

0   1   2   3   4   5   6   7   8   9   10   11   12

Start at 12. Count back 3 and say 11, 10, 9.   12 − 3 = __9__

Try it

Subtract. You can use the number line.

0   1   2   3   4   5   6   7   8   9   10   11   12

1. 7 − 2 = __5__   6 − 1 = ___   11 − 2 = ___

2. 2 − 0 = ___   6 − 3 = ___   10 − 3 = ___

3.
$$
\begin{array}{ccccccc}
3 & 7 & 7 & 6 & 4 & 12 & 9 \\
-1 & -0 & -3 & -2 & -1 & -3 & -2 \\
\hline
\end{array}
$$

4.
$$
\begin{array}{ccccccc}
10 & 8 & 7 & 3 & 5 & 11 & 10 \\
-1 & -3 & -1 & -0 & -1 & -3 & -2 \\
\hline
\end{array}
$$

Sum it Up! How can you count back in order to subtract 8 − 2?

Math at Home: Your child used a number line to subtract 0, 1, 2, or 3 from a number.
Activity: Say a number between 3 and 12. Have your child subtract 1, 2, or 3. Have your child count back to find the difference.

Practice Subtract. You can use the number line.

<----+----+----+----+----+----+----+----+----+----+----+----+----+---->
0   1   2   3   4   5   6   7   8   9   10   11   12

5. 7 − 1 = _6_   11 − 3 = ___   9 − 3 = ___

6. 10 − 2 = ___   9 − 0 = ___   8 − 1 = ___

7. 6 − 2 = ___   8 − 3 = ___   9 − 1 = ___

8. 11 − 2 = ___   9 − 2 = ___   7 − 3 = ___

9.

5	11	10	11	9	12	10
− 3	− 1	− 2	− 2	− 3	− 1	− 3

10.

0	7	8	7	8	9	6
− 0	− 3	− 3	− 0	− 1	− 2	− 1

11.

12	11	7	10	6	8	11
− 3	− 0	− 2	− 1	− 0	− 2	− 3

Problem Solving

12. 5 birds sit in a tree.
One bird flies away.
How many birds are left?

_____ birds

16   sixteen

Name _____

Learn

You can use addition facts to find subtraction facts.

Related facts have the same numbers.

$$4 + 5 = \underline{9}$$

$$9 - 5 = \underline{4}$$

Try it   Add. Then write a related subtraction fact.

1. $4 + 3 = \underline{7}$ 　　　 $\underline{} - \underline{} = \underline{}$

2. $2 + 3 = \underline{}$ 　　　 $\underline{} - \underline{} = \underline{}$

Sum it Up    What subtraction fact can you write for $5 + 7 = 12$?

Math at Home: Your child learned how to use addition facts to subtract.
Activity: Put 12 objects on the table. Cover some of them. Have your child use subtraction to say how many are missing. For example, $12 - 3 = 9$.

You can use addition facts to find subtraction facts.

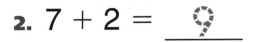

2. 7 + 2 = _9_

_9_ – _7_ = _2_

9 + 1 = ___

___ – ___ = ___

3. 3 + 2 = ___

___ – ___ = ___

6 + 4 = ___

___ – ___ = ___

4. 4 + 5 = ___

___ – ___ = ___

7 + 3 = ___

___ – ___ = ___

Is the addition related to the subtraction?
Write yes or no.

5. 4 + 7 = 11

11 – 4 = 7

_____

4 + 0 = 4

4 – 0 = 4

_____

5 + 2 = 7

5 – 2 = 3

_____

Algebra & functions Missing Addends

Complete each number sentence.

6. 6 + ☐ = 8   8 – ☐ = 6   ☐ – 6 = 2

7. 10 – ☐ = 9   10 – ☐ = 1   ☐ – 9 = 1

8. 4 + ☐ = 4   4 – ☐ = 4   ☐ – 0 = 4

Name _____

Learn

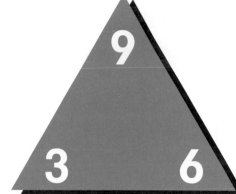

$3 + 6 = 9$ $9 - 3 = 6$

$6 + 3 = 9$ $9 - 6 = 3$

3, 6, and 9 make a fact family.

Try it   Add or subtract. Complete each fact family.

1.

$5 + 2 = \underline{7}$ $7 - 2 = \underline{}$

$2 + 5 = \underline{}$ $7 - 5 = \underline{}$

2.

$4 + \underline{} = 10$ $\underline{} - 4 = 6$

$6 + 4 = \underline{}$ $10 - \underline{} = 4$

3.

$3 + \underline{} = 8$ $8 - 3 = \underline{}$

$5 + 3 = \underline{}$ $\underline{} - 5 = 3$

4.

$5 + \underline{} = 11$ $11 - 5 = \underline{}$

$\underline{} + 5 = 11$ $11 - \underline{} = 5$

   Use the numbers 4, 5, and 9. What fact family can you make?

Math at Home: Your child learned that addition and subtraction facts belong to the same fact family.
Activity: Ask your child to complete the fact family for 9 - 2 = 7.

nineteen   **19**

Add or subtract. Complete each fact family.

5.

$4 + 1 =$ __5__

$1 +$ ___ $= 5$

$5 - 4 =$ ___

___ $- 1 = 4$

6.

$5 + 7 =$ ___

___ $+ 5 = 12$

$12 -$ ___ $= 7$

$12 - 7 =$ ___

7.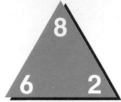

$6 +$ ___ $= 8$

$2 +$ ___ $= 8$

$8 -$ ___ $= 2$

$8 -$ ___ $= 6$

Complete each fact family triangle.

8.

 ___

 ___

 ___

9.

 ___

 ___

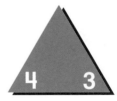 ___

Spiral Review and Test Prep

Choose the correct answer.

10. Which of these does not belong in the fact family?

- ○ $4 + 2 = 6$
- ○ $6 - 4 = 2$
- ○ $8 - 5 = 3$
- ○ $2 + 4 = 6$

11. Which number comes between? | 42 | | 44 |

- ○ 39
- ○ 40
- ○ 41
- ○ 43

Name _____

Subtract.

1. 8 − 5 = _____

2. 11 − 3 = _____

3. 10 − 3 = _____

4. 1 − 1 = _____

5. 8 − 8 = _____

6. 9 − 2 = _____

7. 11 − 4 = _____

8. 5 − 0 = _____

9. 0 − 0 = _____

10. 12 − 7 = _____

11. 3 − 3 = _____

12. 9 − 8 = _____

13. 12 − 6 = _____

14. 10 − 5 = _____

15. 4 − 3 = _____

16. 7 − 1 = _____

17. 12 − 1 = _____

18. 10 − 1 = _____

19. 11 − 3 = _____

20. 9 − 6 = _____

21. 7 − 6 = _____

22. 7 − 3 = _____

23. 6 − 1 = _____

24. 4 − 2 = _____

Math at Home: Your child practiced subtraction facts to sums of 12.
Activity: Cover the answers with a paper strip. Time your child as he or she writes the
answers again. You can repeat daily to help your child recall the facts quickly.

Facts Practice: Addition and Subtraction

1.

7	8	11	4	7	8	12
+0	− 3	− 2	+ 5	+ 3	− 4	− 3

2.

10	3	8	3	12	10	11
− 3	+ 9	− 1	+ 3	− 6	− 6	− 5

3.

7	8	6	7	8	6	10
− 5	+ 2	+ 5	− 2	+ 3	+ 3	− 2

4.

11	9	8	7	10	4	8
− 3	− 7	+ 4	− 6	− 4	+ 3	+ 1

5.

9	4	6	9	2	8	9
− 3	+ 4	− 4	− 5	+7	− 2	+ 0

6.

7	5	9	6	2	11	9
+ 4	− 5	− 7	+ 4	+ 9	− 7	− 0

7.

12	11	5	8	1	10	7
− 2	− 6	+ 5	+ 0	+ 9	− 0	+ 5

Name _____

Draw a Pond

You want to draw a pond picture.
Choose some pond animals and plants.

Plan your pond.
Choose some animals and plants.
Make a list.

_____

_____

_____

_____

_____

_____

_____

_____

_____

Workspace

2. Draw the animals and plants you chose.
Label each part of your picture.

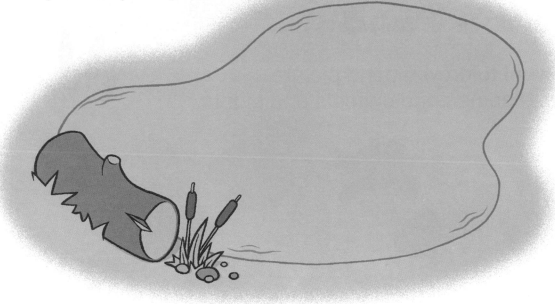

3. Tell about the diagram you made.
How did you decide which parts to label?

_____

_____

_____

4. **What if** you want to add two more animals?
Draw two animals and label them.

Workspace

Name _____

Why Do Ducks Stay Warm?

Ducks stay warm because water rolls off their feathers.

What to do

1. Draw a big feather.

2. Color one half of the feather.

3. Have your partner hold up the paper.

4. Put 2 drops of water on the colored part of the feather.

5. Now put 2 drops of water on the other side of the feather.

What did you find out?

6. What happened to the side of the feather you colored?

_____

7. What happened to side of the feather you did not color?

_____

8. What observations did you make? Explain.

_____

_____

Want to do more?

How do you think feathers help a duck stay warm?

Name _____

Subtract. You can use the number line.

0   1   2   3   4   5   6   7   8   9   10   11   12

1. 10 − 2 = ____   12 − 3 = ____

2. 9 − 1 = ____   11 − 2 = ____

Add. Then write a related subtraction fact.

3. 6 + 6 = ____   9 + 2 = ____

____ − ____ = ____   ____ − ____ = ____

4. 10 + 0 = ____   3 + 3 = ____

____ − ____ = ____   ____ − ____ = ____

Complete the fact family.

5.

5 + 7 = ____   12 − 5 = ____

7 + 5 = ____   12 − 7 = ____

Draw a picture. Solve.

6. There are 5 ducks on the beach.
3 ducks swim away.
How many ducks are on the
beach now?

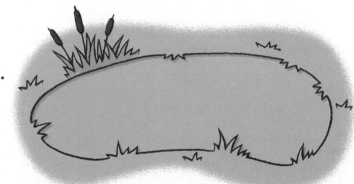

_____ ducks

Name_____

At the Pond

You Will Need

● counter

[5] number cube

- Your and a partner take turns.

- Put your counter on start.

- Roll a [5]. Move that many spaces.

- Find the sum or difference. Have your partner check the answer.

- If correct, stay in the space. If not, move back 1 space.

- Play until you reach Lookout Beach.

Start

6 + 3

10 − 3

7 − 3

1 + 4

5 + 5

9 − 2

3 + 3

3 − 1

7 + 1

9 − 1

9 + 3

8 − 2

6 − 2

6 + 6

2 + 9

8 − 4

9 − 5

7 + 5

6 + 2

4 + 4

Lookout Beach

Chapter Review

Name _____

Language and Math

Complete.
Use a word from the list.

Math Words

addends
sum
difference
fact family

1. The _____ of $5 + 2$ is 7.

Read these words.

2. You can use the numbers 2, 6, and 8

   to make a _____.

3. The _____ of $9 - 3$ is 6.

4. In $5 + 6 = 11$, 5 and 6 are _____.

Concepts and Skills

Add or subtract.

5. $7 + 2 = $ ____ $5 + 3 = $ ____

6. $7 - 2 = $ ____ $10 - 1 = $ ____

7.
$$\begin{array}{cc} 4 \\ +\,4 \\ \hline \end{array} \qquad \begin{array}{cc} 4 \\ +\,5 \\ \hline \end{array} \qquad \begin{array}{cc} 7 \\ -\,1 \\ \hline \end{array} \qquad \begin{array}{cc} 5 \\ +\,3 \\ \hline \end{array} \qquad \begin{array}{cc} 3 \\ +\,5 \\ \hline \end{array}$$

Add or subtract.

8.
$$5 + 5$$ $$10 - 5$$ $$3 + 3$$ $$6 - 3$$ $$7 - 0$$

9.
$$4 - 1$$ $$10 + 1$$ $$8 + 3$$ $$6 + 6$$ $$9 - 5$$

Problem Solving

Draw a picture. Solve.

10. 8 frogs were in the pond.
3 more jump in.
How many frogs are there now?

_____ frogs

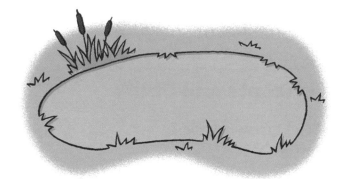

11. There were 3 birds in the tree.
5 more birds came to the tree.
How many birds are there now?

_____ birds

Journal

Tell how you found your answer.

_____

Extra Practice

Name _____

Add or subtract.

| 2 | 5 | 7 | 8 | 10 |

$4 + 3 =$ _____

$6 + 1 =$ _____

$5 - 0 =$ _____

$9 - 2 =$ _____

$6 + 2 =$ _____

$$\begin{array}{r} 8 \\ -\ 3 \\ \hline \end{array}$$

$$\begin{array}{r} 1 \\ +\ 6 \\ \hline \end{array}$$

$5 + 5 =$ _____

$11 - 4 =$ _____

$$\begin{array}{r} 10 \\ -\ 5 \\ \hline \end{array}$$

$10 - 3 =$ _____

$$\begin{array}{r} 3 \\ +\ 7 \\ \hline \end{array}$$

$$\begin{array}{r} 11 \\ -\ 6 \\ \hline \end{array}$$

$11 - 3 =$ _____

$$\begin{array}{r} 7 \\ -\ 2 \\ \hline \end{array}$$

$8 + 0 =$ _____

$10 - 8 =$ _____

Look at the function table. What could the addition pattern be? Explain.

In	1	2	3	4	5	6
Out	4	5	6	7	8	9

The rule is __add 3__ to each number in the first row to get each number in the next row.

Find a rule that describes a pattern for each function table. Then complete the table.

1.

In	1	2	3	4	5	6
Out	3	4	5	6		

The rule is _____.

2.

In	12	10	8	6	4	2
Out	10	8	6			0

The rule is _____.

3.

In	1	2	3	4	5	6
Out	6			9	10	11

The rule is _____.

4.

In	12	11	10	9	8	7
Out	5	4			1	

The rule is _____.

Critical Thinking

5. How does the *In* number change in table 2?

_____

6. How does the *Out* number change in table 2?

_____

Chapter Test

Name _____

Add.

1.
$$\begin{array}{r} 4 \\ +\ 4 \\ \hline \end{array}$$
$$\begin{array}{r} 4 \\ +\ 5 \\ \hline \end{array}$$
$$\begin{array}{r} 8 \\ +\ 4 \\ \hline \end{array}$$
$$\begin{array}{r} 4 \\ +\ 8 \\ \hline \end{array}$$
$$\begin{array}{r} 5 \\ +\ 6 \\ \hline \end{array}$$
$$\begin{array}{r} 6 \\ +\ 5 \\ \hline \end{array}$$

Subtract.

2.
$$\begin{array}{r} 8 \\ -\ 2 \\ \hline \end{array}$$
$$\begin{array}{r} 8 \\ -\ 4 \\ \hline \end{array}$$
$$\begin{array}{r} 7 \\ -\ 4 \\ \hline \end{array}$$
$$\begin{array}{r} 7 \\ -\ 3 \\ \hline \end{array}$$
$$\begin{array}{r} 10 \\ -\ 5 \\ \hline \end{array}$$
$$\begin{array}{r} 10 \\ -\ 4 \\ \hline \end{array}$$

Add or subtract.

3. $6 + 6 =$ _____ $12 - 6 =$ _____ $6 - 3 =$ _____

4. $7 - 5 =$ _____ $10 + 2 =$ _____ $6 - 1 =$ _____

Draw a picture. Solve.

5. 8 frogs were in the pond. 2 more jump in. How many frogs are there now?

_____ frogs

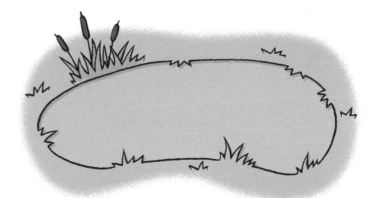

6. 4 turtles were on a log. 3 more turtles get on the log. How many turtles are on the log now?

_____ turtles

Use the information in the picture.

1. Write an addition problem. Solve.

_____

_____

2. Write a subtraction problem. Solve.

_____

_____

 You may want to put this page in your portfolio.

Name _____

Choose the correct answer.

Number Sense

1. Add.

$$5 + 5$$

○ 0
○ 2
○ 5
○ 10

2. Subtract.

$$8 - 3$$

○ 3
○ 5
○ 8
○ 11

3. Which of these is a doubles fact?

○ 3 + 3
○ 3 + 4
○ 3 + 5
○ 3 + 6

Algebra and Functions

4. Which number sentence can you use to solve this problem?

Bill has 5 hats. Jo has 3 hats. How many hats do they have altogether?

○ 5 + 3 = 8
○ 5 − 3 = 2
○ 5 − 4 = 1
○ 3 + 8 = 11

5. Which sign belongs in the box?

7 ☐ 2 = 5

○ 5
○ +
○ −
○ ×

6. 4 frogs are at the pond. 3 hop away. Which shows how many frogs are left?

○ 3 + 4 = 7
○ 4 − 1 = 3
○ 4 − 3 = 1
○ 7 + 3 = 10

McGraw-Hill School Division

Measurement and Geometry

7. Which figure shows a circle?

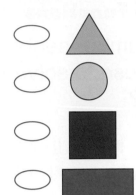

- ⬭
- ⬭
- ⬭
- ⬭

8. Which day comes just after Thursday?

- ⬭ Tuesday
- ⬭ Wednesday
- ⬭ Thursday
- ⬭ Friday

9. Which string is longer than the key?

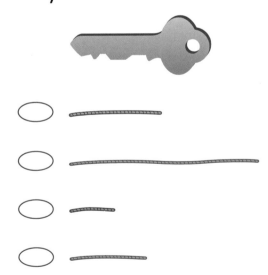

- ⬭
- ⬭
- ⬭
- ⬭

Mathematical Reasoning

10. 6 flowers float on lily pads. 3 more flowers bloom. How many flowers are on the lily pads now?

- ⬭ 9
- ⬭ 3
- ⬭ 10
- ⬭ 6

11. The beaver uses 11 logs to build a dam. Then it uses 1 more log. How many logs does it use in all?

- ⬭ 10
- ⬭ 11
- ⬭ 9
- ⬭ 12

12. There are 9 fish swimming in the pond. 5 fish swim away. How many fish are in the pond now?

- ⬭ 6
- ⬭ 4
- ⬭ 3
- ⬭ 5

Explain how you found out.

_____

Addition and Subtraction Strategies and Facts to 20

theme
Postcards from Pete

Use the Data
Look at the picture.
Tell a number story.

Hi!
It's me Pete.
The Grand Canyon
is big! I like
the donkeys best.
Love,
Pete

To:
Sam Harper
32 Maple Street
Franklin, MA 02038

What You Will Learn
In this chapter you will learn how to:

• Add, facts to 20.

• Subtract, facts to 20.

• Add three numbers.

• Write number sentences to solve problems.

Hi!
It's me,
Pete!

Story by
Joe Rubino
Illustrated by
Bruce Van Patter

Math Words

sum

6 + 8 = 14

sum

difference

17 − 9 = 8

difference

doubles

8 + 8 = 16

fact family

4 + 3 = 7   7 − 3 = 4
3 + 4 = 7   7 − 4 = 3

Dear Family,

In Chapter 2, I will learn strategies for adding and subtracting numbers to 20. Here are new vocabulary words and an activity that we can do together.

Seeing Double

• Put from 9 bottle caps in a pile.

use

bottle caps

• Have your child count out the same number of bottle caps.

• Ask your child to tell you the total number of bottle caps.

• Repeat the activity using different numbers of bottle caps.

Additional activities at www.mhschool.com/math

Name _____ **Doubles to Add**

Learn

You can use doubles to find a sum.

7 + 7 is 14.
7 + 8 is 1 more.
So 7 + 8 is 15.

7 + 7 = __14__   7 + 8 = __15__

Try it   Find each sum.

1. 8 + 8 = __16__   8 + 9 = __17__   8 + 7 = __15__

2. 9 + 9 = ___   9 + 10 = ___   9 + 8 = ___

3.
6	6¢	6	8¢	8	8	4
+ 6	+ 7¢	+ 5	+ 8¢	+ 9	+ 7	+ 5

4.
7	7	7¢	9	9¢	9	5
+ 7	+ 8	+ 6¢	+ 9	+ 10¢	+ 8	+ 5

Sum it Up    How can the doubles fact 8 + 8 = 16 help you find 8 + 9 and 8 + 7?

 Math at Home: Your child used doubles to add.
Activity: Have your child draw 4 spiders and then tell you a doubles story about the picture.

McGraw-Hill School Division

Practice Add.

5. $7 + 7 = \underline{14}$ $7 + 8 = \underline{\hspace{1cm}}$ $7 + 6 = \underline{\hspace{1cm}}$

6. $8 + 8 = \underline{\hspace{1cm}}$ $8 + 9 = \underline{\hspace{1cm}}$ $8 + 7 = \underline{\hspace{1cm}}$

7. $9 + 9 = \underline{\hspace{1cm}}$ $9 + 10 = \underline{\hspace{1cm}}$ $9 + 8 = \underline{\hspace{1cm}}$

8.

9	8	5¢	7	6	7¢	6
+ 0	+ 2	+ 6¢	+ 8	+ 7	+ 7¢	+ 6

9.

8¢	8	8	6¢	6	7	5
+ 8¢	+ 9	+ 7	+ 6¢	+ 5	+ 6	+ 5

Problem Solving

Number Sense

10. Pete has 7 New Mexico state stamps. Toby has the same number of stamps from Ohio. How many stamps do they have in all?

_____ stamps

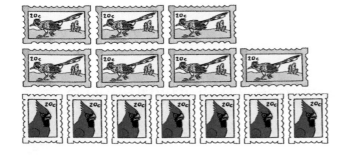

11. There is 1 more stamp from Kansas than from New Mexico. There are 7 stamps from New Mexico. How many stamps are there altogether?

_____ stamps

Name _____

Learn

Add 8 + 6.
Start with 8. Add 6.

You can make a ten to help you add.

You made a 10 and have 4 extra. 8 + 6 is the same as 10 + 4

10 + 4 = __14__

8 + 6 = __14__

Try it

Add. You can use 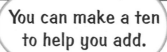 and the 10-frame.

1. 7 + 4 = __11__   9 + 3 = ____   8 + 4 = ____

2.
$\begin{array}{r} 7 \\ +5 \\ \hline \end{array}$
$\begin{array}{r} 8 \\ +6 \\ \hline \end{array}$
$\begin{array}{r} 9 \\ +4 \\ \hline \end{array}$
$\begin{array}{r} 7 \\ +3 \\ \hline \end{array}$
$\begin{array}{r} 8 \\ +3 \\ \hline \end{array}$
$\begin{array}{r} 9 \\ +2 \\ \hline \end{array}$
$\begin{array}{r} 6 \\ +6 \\ \hline \end{array}$

3.
$\begin{array}{r} 9 \\ +5 \\ \hline \end{array}$
$\begin{array}{r} 7 \\ +6 \\ \hline \end{array}$
$\begin{array}{r} 8 \\ +2 \\ \hline \end{array}$
$\begin{array}{r} 8 \\ +6 \\ \hline \end{array}$
$\begin{array}{r} 7 \\ +7 \\ \hline \end{array}$
$\begin{array}{r} 9 \\ +6 \\ \hline \end{array}$
$\begin{array}{r} 9 \\ +1 \\ \hline \end{array}$

Sum it Up

How can making a 10 help you add?

Math at Home: Your child used a 10-frame to add 7, 8, and 9.
Activity: Have your child show you how to add using the 10-frame and some of the problems on this page.

Practice Add. You can use  and the 10-frame.

4. $9 + 7 =$ __16__

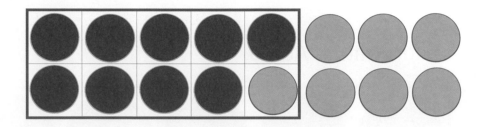

5. $5 + 8 =$ _____ $4 + 8 =$ _____ $4 + 7 =$ _____

6. $5 + 9 =$ _____ $7 + 9 =$ _____ $8 + 7 =$ _____

7.
$$\begin{array}{cccccc} 9 & 8¢ & 6 & 3 & 3¢ & 4¢ & 6 \\ +\,6 & +\,6¢ & +\,9 & +\,8 & +\,9¢ & +\,9¢ & +\,8 \end{array}$$

8.
$$\begin{array}{ccccccc} 7 & 9 & 7¢ & 8 & 9 & 5¢ & 5 \\ +\,4 & +\,3 & +\,5¢ & +\,9 & +\,4 & +\,7¢ & +\,9 \end{array}$$

Critical Thinking Journal

9. Read the card.
How many animals did Pete see? Explain.

_____ animals

_____

_____

I saw 5 seals and 5 sea lions. They were cute!

Love, Pete

Grandpa Clark

P.O. Box 4

Mellin Village, NH

03850

Learn

You can use different strategies to add.

I added 4 + 6 first to make 10. Then I added 10 + 4.

```
  4 ▸ 10
  6
+ 4
 14
```

I added the doubles 4 + 4 first to get 8. Then I added 8 + 6.

```
  4
  6 ▸ 8
+ 4
 14
```

Try it   Find each sum.

1.
```
  8 ▸ 10
  2
+ 3
 13
```
```
  2
  5 ▸ ☐
+ 5
```
```
  8
  1 ▸ ☐
+ 9
```
```
  7 ▸ ☐
  7
+ 2
```

2.
```
  7        4        1        3        7        7        8
  7        7        6        3        3        3        2
+ 2      + 3      + 4      + 5      + 2      + 5      + 7
```

3. 8 + 3 + 3 = ____   9 + 1 + 5 = ____

What strategy would you use to find the sum of 3 + 7 + 2?

Math at Home: Your child added three numbers by using doubles and making tens.
Activity: Activity: Have your child show you how to add 6 + 6 + 7.

Practice   Add. Use different stategies.

4.

6	8	5	4	6	4	6
3	2	5	3	6	5	4
+ 6	+ 4	+ 1	+ 7	+ 3	+ 6	+ 2

15

5.

4	7	6	8	8	5	3
6	7	6	8	3	6	3
+ 7	+ 2	+ 5	+ 1	+ 2	+ 4	+ 4

6.

8	9	8	6	2	3	2
1	2	2	5	8	3	9
+ 9	+ 1	+ 7	+ 4	+ 6	+ 8	+ 0

7. $4 + 6 + 7 =$ _____ $5 + 8 + 5 =$ _____

8. $8 + 8 + 2 =$ _____ $9 + 7 + 3 =$ _____

Spiral Review and Test Prep

Choose the correct answer.

9. $2 + 8 + 5 = \boxed{}$

- ◯ 10
- ◯ 13
- ◯ 15
- ◯ 17

10. $6 + 7 + 6 = \boxed{}$

- ◯ 12
- ◯ 13
- ◯ 17
- ◯ 19

Name _____

Add.

1. $7 + 7 =$ _____

2. $7 + 8 =$ _____

3. $9 + 5 =$ _____

4. $8 + 8 =$ _____

5. $7 + 9 =$ _____

6. $12 + 0 =$ _____

7. $9 + 4 =$ _____

8. $6 + 8 =$ _____

9. $8 + 9 =$ _____

10. $5 + 8 =$ _____

11. $8 + 5 =$ _____

12. $7 + 6 =$ _____

13. $6 + 9 =$ _____

14. $9 + 7 =$ _____

15. $8 + 6 =$ _____

16. $4 + 9 =$ _____

17. $6 + 7 =$ _____

18. $9 + 8 =$ _____

19. $10 + 10 =$ _____

20. $10 + 9 =$ _____

21. $6 + 6 =$ _____

22. $9 + 9 =$ _____

23. $4 + 9 =$ _____

24. $8 + 7 =$ _____

Math at Home: Your child practiced addition facts.
Activity: Cover the answers with a paper strip. Time your child as he or she writes
the answers again. You can repeat daily to help your child recall the facts quickly.

forty-five **45**

Facts Practice: Addition and Subtraction

Add or subtract.

1.
$$\begin{array}{r} 7 \\ + 7 \\ \hline \end{array} \qquad \begin{array}{r} 4 \\ + 9 \\ \hline \end{array} \qquad \begin{array}{r} 12 \\ - 6 \\ \hline \end{array} \qquad \begin{array}{r} 11 \\ - 5 \\ \hline \end{array} \qquad \begin{array}{r} 5 \\ + 8 \\ \hline \end{array} \qquad \begin{array}{r} 10 \\ - 7 \\ \hline \end{array} \qquad \begin{array}{r} 6 \\ + 6 \\ \hline \end{array}$$

2.
$$\begin{array}{r} 6 \\ + 7 \\ \hline \end{array} \qquad \begin{array}{r} 7 \\ - 4 \\ \hline \end{array} \qquad \begin{array}{r} 8 \\ + 7 \\ \hline \end{array} \qquad \begin{array}{r} 9 \\ - 5 \\ \hline \end{array} \qquad \begin{array}{r} 12 \\ - 9 \\ \hline \end{array} \qquad \begin{array}{r} 6 \\ + 8 \\ \hline \end{array} \qquad \begin{array}{r} 10 \\ - 5 \\ \hline \end{array}$$

3.
$$\begin{array}{r} 9 \\ - 9 \\ \hline \end{array} \qquad \begin{array}{r} 5 \\ + 9 \\ \hline \end{array} \qquad \begin{array}{r} 11 \\ - 9 \\ \hline \end{array} \qquad \begin{array}{r} 10 \\ - 6 \\ \hline \end{array} \qquad \begin{array}{r} 8 \\ + 8 \\ \hline \end{array} \qquad \begin{array}{r} 11 \\ - 7 \\ \hline \end{array} \qquad \begin{array}{r} 12 \\ - 6 \\ \hline \end{array}$$

4.
$$\begin{array}{r} 6 \\ + 9 \\ \hline \end{array} \qquad \begin{array}{r} 10 \\ + 6 \\ \hline \end{array} \qquad \begin{array}{r} 8 \\ + 5 \\ \hline \end{array} \qquad \begin{array}{r} 12 \\ - 8 \\ \hline \end{array} \qquad \begin{array}{r} 7 \\ + 8 \\ \hline \end{array} \qquad \begin{array}{r} 8 \\ - 4 \\ \hline \end{array} \qquad \begin{array}{r} 5 \\ + 9 \\ \hline \end{array}$$

5.
$$\begin{array}{r} 10 \\ - 2 \\ \hline \end{array} \qquad \begin{array}{r} 8 \\ + 9 \\ \hline \end{array} \qquad \begin{array}{r} 10 \\ - 7 \\ \hline \end{array} \qquad \begin{array}{r} 8 \\ + 6 \\ \hline \end{array} \qquad \begin{array}{r} 11 \\ - 8 \\ \hline \end{array} \qquad \begin{array}{r} 9 \\ + 9 \\ \hline \end{array} \qquad \begin{array}{r} 10 \\ - 3 \\ \hline \end{array}$$

6.
$$\begin{array}{r} 8 \\ - 8 \\ \hline \end{array} \qquad \begin{array}{r} 10 \\ + 10 \\ \hline \end{array} \qquad \begin{array}{r} 7 \\ + 9 \\ \hline \end{array} \qquad \begin{array}{r} 10 \\ - 10 \\ \hline \end{array} \qquad \begin{array}{r} 8 \\ - 3 \\ \hline \end{array} \qquad \begin{array}{r} 10 \\ - 8 \\ \hline \end{array} \qquad \begin{array}{r} 8 \\ + 5 \\ \hline \end{array}$$

7.
$$\begin{array}{r} 10 \\ + 9 \\ \hline \end{array} \qquad \begin{array}{r} 8 \\ + 8 \\ \hline \end{array} \qquad \begin{array}{r} 9 \\ + 6 \\ \hline \end{array} \qquad \begin{array}{r} 7 \\ - 4 \\ \hline \end{array} \qquad \begin{array}{r} 16 \\ - 8 \\ \hline \end{array} \qquad \begin{array}{r} 7 \\ - 7 \\ \hline \end{array} \qquad \begin{array}{r} 2 \\ + 10 \\ \hline \end{array}$$

Name _____

Use a Summary

Reading Skill   You can use a summary to help you solve a story problem

Grandpa and I found baby ducks today. There were 4 baby ducks in the pond. There were 6 more baby ducks asleep in the nest.

Solve.

1. Write 2 important number facts from the story.

_____   _____

2. Use the facts to write a sentence that tells a number story.

_____

3. How many ducks did Pete see? _____ ducks

Write a number sentence to show your thinking.

_____

Math at Home: Your child read a story and found a short way to summarize it.
Activity: Tell your child a simple number story and then have your child summarize it.

Solve.

Grandpa and I went fishing at the lake. I caught 8 fish today. I have 3 crickets in my cricket jar.

Solve.

4. Write 2 important number facts from the story. _____

_____

5. Use the facts to write a sentence that tells a number story.

_____

6. How many animals did Pete see?
Write a number sentence to show your thinking.

_____

Journal

7. Use the picture. Write a problem.

_____

Name _____

Add.

1. 7 + 7 = ____   7 + 8 = ____   6 + 6 = ____

2. 6 + 7 = ____   9 + 7 = ____   7 + 6 = ____

3. 9 + 9 = ____   9 + 8 = ____   8 + 8 = ____

4. 8 + 7 = ____   5 + 9 = ____   10 + 10 = ____

5.

7	8¢	9	8	3¢	6¢
+ 4	+ 4¢	+ 2	+ 3	+ 9¢	+ 8¢

6.

7	4	7	1	6	9
3	4	7	5	6	9
+ 6	+ 9	+ 2	+ 9	+ 5	+ 2

Journal

7. What strategy would you use first to add 8 + 2 + 4?

_____

Add.

1.

$$\begin{array}{r} 5 \\ +\ 1 \\ \hline \end{array} \qquad \begin{array}{r} 8 \\ +\ 4 \\ \hline \end{array} \qquad \begin{array}{r} 4 \\ +\ 3 \\ \hline \end{array} \qquad \begin{array}{r} 5 \\ +\ 5 \\ \hline \end{array} \qquad \begin{array}{r} 6 \\ +\ 3 \\ \hline \end{array} \qquad \begin{array}{r} 7 \\ +\ 1 \\ \hline \end{array} \qquad \begin{array}{r} 6 \\ +\ 6 \\ \hline \end{array}$$

Subtract.

2.

$$\begin{array}{r} 11 \\ -\ 3 \\ \hline \end{array} \qquad \begin{array}{r} 10 \\ -\ 3 \\ \hline \end{array} \qquad \begin{array}{r} 8 \\ -\ 4 \\ \hline \end{array} \qquad \begin{array}{r} 12 \\ -\ 6 \\ \hline \end{array} \qquad \begin{array}{r} 9 \\ -\ 0 \\ \hline \end{array} \qquad \begin{array}{r} 7 \\ -\ 3 \\ \hline \end{array} \qquad \begin{array}{r} 11 \\ -\ 5 \\ \hline \end{array}$$

Write each number.

3. two _____   seven _____   eight _____

TECHNOLOGY LINK

Use a Table to Add

- Use a table.
- Choose a mat to show words and numbers. Add 2 rows.
- Enter Addend 3 times. Enter 4, 6, and 7.
- Click the table key to find total in last row.
- What is the sum? Enter other numbers. Find the sum.

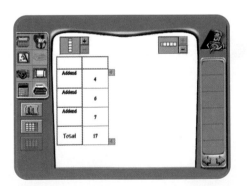

For more practice use Math Traveler™.

Name _____

Write a Number Sentence

Read ▶ Jan has 8 postcards.
Carlos has 4 postcards.
How many postcards
do Jan and Carlos
have altogether?

What do I know?

What do I need
to find out?

Plan ▶ I can write a number sentence to solve.

Solve ▶ $8 + 4 = 12$

Jan and Carlos have 12 postcards.

Look Back ▶ Does my answer make sense? Why?

Write a number sentence. Solve.

1. Lyn has 6 stamps from Mexico. She has
the same number of stamps from Italy.
How many stamps does Lyn have in all?

_____ ◯ _____ = _____

_____ stamps

Which words helped you decide whether to add or
subtract?

Math at Home: Your child solved problems by writing addition and
subtraction sentences.
Activity: Tell a simple addition or subtraction problem for your child to solve.

Write a number sentence. Solve.

2. Jenny has 9 postcard stamps and 3 letter stamps. How many more postcard stamps than letter stamps does she have?

_____ ◯ _____ = _____

_____ postcard stamps

3. David has 7 air mail stamps and 6 letter stamps. How many stamps does he have altogether?

_____ ◯ _____ = _____

_____ stamps

4. Sherrie buys 2 bird stamps. Each stamp cost 10¢. How much does Sherrie spend?

_____ ◯ _____ = _____

_____ ¢

5. Ben buys a stamp that cost 8¢. He buys another stamp that costs 9¢. How much do the stamps cost?

_____ ◯ _____ = _____

_____ ¢

 **Critical Thinking**

 Journal

6. Look at the stamps.
Use numbers.
Write a problem.

_____

Learn

Doubles can help you subtract 16 − 8.

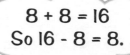

$8 + 8 = 16$
So $16 - 8 = 8$.

$8 + 8 = \underline{16}$ $16 - 8 = \underline{8}$

Try it   Add or subtract.

1. $7 + 7 = \underline{14}$ $14 - 7 = \underline{7}$

2. $9 + 9 = \underline{\hphantom{00}}$ $18 - 9 = \underline{\hphantom{00}}$

3.
$$\begin{array}{r} 6 \\ +\ 6 \\ \hline \end{array} \qquad \begin{array}{r} 12 \\ -\ 6 \\ \hline \end{array} \qquad \begin{array}{r} 7 \\ +\ 7 \\ \hline \end{array} \qquad \begin{array}{r} 14 \\ -\ 7 \\ \hline \end{array} \qquad \begin{array}{r} 10 \\ +\ 10 \\ \hline \end{array} \qquad \begin{array}{r} 20 \\ -\ 10 \\ \hline \end{array}$$

4.
$$\begin{array}{r} 9 \\ +\ 9 \\ \hline \end{array} \qquad \begin{array}{r} 18 \\ -\ 9 \\ \hline \end{array} \qquad \begin{array}{r} 8 \\ +\ 8 \\ \hline \end{array} \qquad \begin{array}{r} 16 \\ -\ 8 \\ \hline \end{array} \qquad \begin{array}{r} 5 \\ +\ 5 \\ \hline \end{array} \qquad \begin{array}{r} 10 \\ -\ 5 \\ \hline \end{array}$$

Sum it Up   What doubles fact can you use to solve $14 - 7$?

Math at Home: Your child used addition doubles like 7 + 7 = 14 to subtract.
Activity: Give your child a doubles fact such as 8 + 8. Then have your child use the
fact to tell a related subtraction fact.

McGraw-Hill School Division

Doubles can help you subtract.

5. $8 + 8 = \underline{16}$ $16 - 8 = \underline{}$

6. $9 + 9 = \underline{}$ $18 - 9 = \underline{}$

7. $7 + 7 = \underline{}$ $14 - 7 = \underline{}$

8. $6 + 6 = \underline{}$ $12 - 6 = \underline{}$

9.
| $\begin{array}{r} 6 \\ +6 \\ \hline \end{array}$ | $\begin{array}{r} 12 \\ -6 \\ \hline \end{array}$ | $\begin{array}{r} 7 \\ +7 \\ \hline \end{array}$ | $\begin{array}{r} 14 \\ -7 \\ \hline \end{array}$ | $\begin{array}{r} 5 \\ +5 \\ \hline \end{array}$ | $\begin{array}{r} 10 \\ -5 \\ \hline \end{array}$ |

10.
| $\begin{array}{r} 9 \\ +9 \\ \hline \end{array}$ | $\begin{array}{r} 18 \\ -9 \\ \hline \end{array}$ | $\begin{array}{r} 10 \\ +10 \\ \hline \end{array}$ | $\begin{array}{r} 20 \\ -10 \\ \hline \end{array}$ | $\begin{array}{r} 8 \\ +8 \\ \hline \end{array}$ | $\begin{array}{r} 16 \\ -8 \\ \hline \end{array}$ |

Critical Thinking   Journal

11. What could the pattern be?

Based on your pattern, what is the next number? Explain.

Name _____

Learn

Johnny wrote these related facts.

You can use addition facts to subtract.

9+7=16
16-9=7

MAIL

Try it   Complete each number sentence.

1. 8 + 6 = __14__   7 + 9 = ____   9 + 6 = ____

14 − 8 = __6__   16 − 7 = ____   15 − 9 = ____

2. 8 + 9 = ____   6 + 7 = ____   8 + 8 = ____

17 − 8 = ____   13 − 6 = ____   16 − 8 = ____

3.
$$\begin{array}{r} 8 \\ + 7 \\ \hline \end{array}$$
$$\begin{array}{r} 15 \\ - 8 \\ \hline \end{array}$$
$$\begin{array}{r} 9 \\ + 5 \\ \hline \end{array}$$
$$\begin{array}{r} 14 \\ - 9 \\ \hline \end{array}$$
$$\begin{array}{r} 7 \\ + 7 \\ \hline \end{array}$$
$$\begin{array}{r} 14 \\ - 7 \\ \hline \end{array}$$

4.
$$\begin{array}{r} 5 \\ + 8 \\ \hline \end{array}$$
$$\begin{array}{r} 13 \\ - 5 \\ \hline \end{array}$$
$$\begin{array}{r} 9 \\ + 10 \\ \hline \end{array}$$
$$\begin{array}{r} 19 \\ - 9 \\ \hline \end{array}$$
$$\begin{array}{r} 4 \\ + 9 \\ \hline \end{array}$$
$$\begin{array}{r} 13 \\ - 4 \\ \hline \end{array}$$

Sum it Up

What subtraction fact relates to 7 + 8 = 15?

Math at Home: Your learned about related subtraction and addition facts, like 9 + 8 = 17 and 17 - 9 = 8.
Activity: Ask your child to explain how 9 + 8 = 17 helps to find 17 - 9.

Complete each number sentence.

5. 5 + 8 = __13__   8 + 7 = ____   9 + 4 = ____

13 − 5 = __8__   15 − 8 = ____   13 − 9 = ____

6. 5 + 9 = ____   9 + 10 = ____   7 + 7 = ____

14 − 5 = ____   19 − 9 = ____   14 − 7 = ____

7.
$\begin{array}{r} 9 \\ + 7 \\ \hline \end{array}$
$\begin{array}{r} 16 \\ - 9 \\ \hline \end{array}$
$\begin{array}{r} 6 \\ + 7 \\ \hline \end{array}$
$\begin{array}{r} 13 \\ - 6 \\ \hline \end{array}$
$\begin{array}{r} 9 \\ + 8 \\ \hline \end{array}$
$\begin{array}{r} 17 \\ - 9 \\ \hline \end{array}$

8.
$\begin{array}{r} 8 \\ + 6 \\ \hline \end{array}$
$\begin{array}{r} 14 \\ - 8 \\ \hline \end{array}$
$\begin{array}{r} 9 \\ + 9 \\ \hline \end{array}$
$\begin{array}{r} 18 \\ - 9 \\ \hline \end{array}$
$\begin{array}{r} 6 \\ + 9 \\ \hline \end{array}$
$\begin{array}{r} 15 \\ - 6 \\ \hline \end{array}$

Problem Solving

Use Data

Use the graph to solve.

9. How many bird stamps and tree stamps are there?

_____ stamps

10. How many more bird stamps are there than tree stamps?

_____ more bird stamps

Stamps

Each picture stands for 1 stamp.

Learn

Find the missing addend.

Use a related subtraction fact to help you find the missing number.

$$9 + \boxed{} = 14$$

$$14 - 9 = 5$$

$$So, 9 + 5 = 14$$

Try it Find each missing addend.

1. $5 + \boxed{7} = 12$ $12 - 5 = \underline{7}$

2. $8 + \boxed{} = 17$ $17 - 8 = \boxed{}$

3. $9 + \boxed{} = 13$ $13 - 9 = \boxed{}$

4.
$$\begin{array}{r} 7 \\ + \boxed{} \\ \hline 15 \end{array} \qquad \begin{array}{r} 15 \\ - 7 \\ \hline \end{array}$$

$$\begin{array}{r} 6 \\ + \boxed{} \\ \hline 13 \end{array} \qquad \begin{array}{r} 13 \\ - 6 \\ \hline \end{array}$$

$$\begin{array}{r} 9 \\ + \boxed{} \\ \hline 15 \end{array} \qquad \begin{array}{r} 15 \\ - 9 \\ \hline \end{array}$$

Sum it Up! How does using a related subtraction fact help you find the missing number?

🏠 **Math at Home:** Your child used related subtraction facts to find missing numbers.
Activity: Ask your child to tell you the subtraction fact that will help him or her add
$9 + \boxed{} = 16$.

Find each missing addend.

5. $4 + \boxed{9} = 13$ $13 - 4 = \boxed{}$

6. $7 + \boxed{} = 15$ $15 - 7 = \boxed{}$

7.
$$\begin{array}{r} 6 \\ + \boxed{} \\ \hline 15 \end{array} \qquad \begin{array}{r} 15 \\ - 6 \\ \hline \end{array}$$

$$\begin{array}{r} 8 \\ + \boxed{} \\ \hline 13 \end{array} \qquad \begin{array}{r} 13 \\ - 8 \\ \hline \end{array}$$

$$\begin{array}{r} 9 \\ + \boxed{} \\ \hline 16 \end{array} \qquad \begin{array}{r} 16 \\ - 9 \\ \hline \end{array}$$

Complete each number sentence.

8. $9 + \boxed{} = 16$ $8 + \boxed{} = 14$ $\boxed{} + 6 = 15$

9. $4 \bigcirc 9 = 13$ $14 \bigcirc 5 = 9$ $6 \bigcirc 7 = 13$

10. $20 \bigcirc 10 = 10$ $16 \bigcirc 8 = 8$ $9 \bigcirc 8 = 17$

Find the doubles.

11. $\boxed{} + \boxed{} = 12$ $\boxed{} + \boxed{} = 16$

12. $\boxed{} + \boxed{} = 20$ $\boxed{} + \boxed{} = 18$

13. $\boxed{} + \boxed{} = 14$ $\boxed{} + \boxed{} = 10$

Fact Families

Algebra
& functions

Learn

The numbers 8, 7, and 15 make up this fact family.

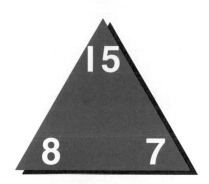

$8 + 7 = \underline{15}$

$7 + 8 = \underline{15}$

$15 - 7 = \underline{8}$

$15 - 8 = \underline{7}$

You can use facts you know to find other facts.

Try it   Complete each fact family.

1.

$9 + 7 = \underline{16}$ $7 + 9 = \underline{}$

$16 - 9 = \underline{}$ $16 - 7 = \underline{}$

2.

$4 + 9 = \underline{}$ $9 + 4 = \underline{}$

$13 - 4 = \underline{}$ $13 - 9 = \underline{}$

3.
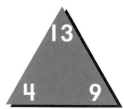

$7 + 7 = \underline{}$ $14 - 7 = \underline{}$

4.

$9 + 8 = \underline{}$ $8 + 9 = \underline{}$

$17 - 9 = \underline{}$ $17 - 8 = \underline{}$

Sum it Up   What fact family can you make with 6, 9, and 15?

Math at Home: Your child learned how related addition and subtraction facts make up a fact family.
Activity: Have your child write a fact family for the numbers 8, 9, and 17.

Complete each fact family.

5.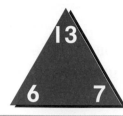

$6 + 7 = \text{13}$

$13 - 6 = \underline{}$

$7 + 6 = \underline{}$

$13 - 7 = \underline{}$

6.

$5 + 9 = \underline{}$

$14 - 5 = \underline{}$

$9 + 5 = \underline{}$

$14 - 9 = \underline{}$

7.

$7 + 8 = \underline{}$

$15 \bigcirc 7 = 8$

$8 + \underline{} = 15$

$15 - \underline{} = 7$

8.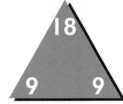

$8 + 9 = \underline{}$

$17 - 8 = \underline{}$

$9 + 8 = \underline{}$

$17 - 9 = \underline{}$

9.

$9 \bigcirc 9 = 18$

$18 \bigcirc 9 = 9$

Spiral Review and Test Prep

Choose the correct answer.

10. Which shows a fact family?

- ⬭ $8 + 8 = 16$ $16 - 8 = 8$
- ⬭ $8 + 8 = 16$ $6 + 6 = 12$
- ⬭ $8 + 8 = 16$ $4 + 4 = 8$
- ⬭ $16 - 8 = 8$ $8 - 4 = 4$

11. Subtract.

$$\begin{array}{r} 12 \\ -2 \\ \hline \end{array}$$

- ⬭ 12
- ⬭ 11
- ⬭ 10
- ⬭ 9

1. $13 - 6 =$ _____

2. $18 - 9 =$ _____

3. $13 - 5 =$ _____

4. $16 - 7 =$ _____

5. $14 - 8 =$ _____

6. $14 - 6 =$ _____

7. $15 - 9 =$ _____

8. $15 - 7 =$ _____

9. $15 - 8 =$ _____

10. $17 - 9 =$ _____

11. $15 - 6 =$ _____

12. $13 - 4 =$ _____

13. $16 - 9 =$ _____

14. $13 - 8 =$ _____

15. $13 - 9 =$ _____

16. $13 - 7 =$ _____

17. $13 - 5 =$ _____

18. $15 - 9 =$ _____

19. $16 - 8 =$ _____

20. $13 - 4 =$ _____

21. $14 - 9 =$ _____

22. $14 - 4 =$ _____

23. $20 - 10 =$ _____

24. $14 - 7 =$ _____

Math at Home: Your child practiced subtraction facts.
Activity: Cover the answers with a paper strip. Time your child as he or she writes the answers again. You can repeat daily to help your child recall the facts quickly.

Facts Practice: Addition and Subtraction

Add or subtract.

1.	8 + 5	11 − 3	8 + 8	11 − 2	6 + 8	8 + 5	9 − 0

2.	9 − 6	7 + 4	11 − 7	12 − 0	9 + 9	10 + 5	3 + 8

3.	10 − 5	6 + 9	12 − 8	7 + 6	10 − 1	11 − 2	12 − 6

4.	5 + 8	12 − 9	12 − 3	6 + 7	9 − 9	7 + 7	11 − 6

5.	11 − 5	8 + 10	8 + 7	9 + 6	12 − 2	10 + 10	9 + 5

6.	4 + 9	8 + 6	8 − 6	5 + 9	12 − 7	10 − 6	7 + 10

7.	8 + 8	7 + 6	12 − 3	11 − 4	10 + 0	8 − 0	4 + 8

Name _____

Design a Stamp!

You want to design a stamp for the state postcard stamp contest.

Contest!
Design a Stamp Card.
What will you show?
A Flower!
A Bird!
A Place!
How much will it cost?
You decide!

You Decide!

Plan how to design your stamp. Choose a picture you want to use. Decide how much the stamp will cost.

_____

_____

_____

_____

_____

Workspace

2. Show what you decided to draw.

3. Tell about how you designed your stamp.
How did you decide how much your stamp cost?

_____

_____

4. What if you decided to
design a letter stamp?
Design a stamp for a
place you like to visit.
Decide how much the
stamp will cost.

Workspace

Name _____

Stamp Glue

Stamps have glue on one side to make them stick.

You Will Need

What to do

1. Label cups A, B, and C.

2. Put 1 spoonful of flour and 2 drops of yellow paint in Cup A.

3. Put 2 spoonfuls of flour and 2 drops of green paint in Cup B.

4. Put 3 spoonfuls of flour and 2 drops of blue paint in Cup C.

5. Fill each cup with water and mix.

6. Use your stamps to try each kind of glue.

Record your findings.

A	
B	
C	

What did you find out?

1. Which glue worked best?
Tell why.

_____

_____

2. How many spoonfuls of flour
did you use in all?

_____ spoonfuls

3. How many drops of paint
did you use in all?

_____ drops

Did You KNOW?

Maple syrup and
tree sap are kinds
of glue.

 Journal

Want to do more?

How could you improve the glue that you made? Try it.

Check Your Progress B

Name _____

Add or subtract.

1.
$$\begin{array}{r} 9 \\ +\ 9 \\ \hline \end{array} \qquad \begin{array}{r} 18 \\ -\ 9 \\ \hline \end{array} \qquad \begin{array}{r} 10 \\ +\ 10 \\ \hline \end{array} \qquad \begin{array}{r} 20 \\ -\ 10 \\ \hline \end{array} \qquad \begin{array}{r} 8 \\ +\ 8 \\ \hline \end{array} \qquad \begin{array}{r} 16 \\ -\ 8 \\ \hline \end{array}$$

Find each missing number.

2. $9 + \underline{\hspace{1cm}} = 16$ $\qquad\qquad$ $16 - 9 = \underline{\hspace{1cm}}$

   $8 + \underline{\hspace{1cm}} = 14$ $\qquad\qquad$ $14 - 8 = \underline{\hspace{1cm}}$

Complete each number sentence.

3. $8 + 8 = \underline{\hspace{1cm}}$ $\qquad\qquad$ $16 - 8 = \underline{\hspace{1cm}}$

   $8 + 7 = \underline{\hspace{1cm}}$ $\qquad\qquad$ $15 - 7 = \underline{\hspace{1cm}}$

Complete each fact family.

4.  $\quad 9 + 6 = \underline{\hspace{1cm}} \qquad 15 - 9 = \underline{\hspace{1cm}}$

   $6 + 9 = \underline{\hspace{1cm}} \qquad 15 - 6 = \underline{\hspace{1cm}}$

Write a number sentence to solve.

5. Jay has 9 airmail stamps and 8 postcard stamps. How many stamps does Jay have?

   $\underline{\hspace{1cm}} + \underline{\hspace{1cm}} = \underline{\hspace{1cm}}$

   _____ stamps

6. Jenny has 10 postcards. She bought 10 more postcards. How many postcards does Jenny have?

   $\underline{\hspace{1cm}} + \underline{\hspace{1cm}} = \underline{\hspace{1cm}}$

   _____ postcards

Name_____

Mail Race

POST OFFICE

Take turns.

- Put your counter on **Start**.

- Roll a number cube.
  Move that many spaces.

- Tell a related addition or subtraction fact.
  Have your partner check.

You Will Need

FINISH

8 + 9
8 + 6

16 – 8	15 – 6	8 + 4	15 – 8	14 – 7	7 + 7

20 – 10	• If **correct**, stay in the space. If **wrong**, move back one space. • The winner is the first player to reach End.

16 – 8	9 + 5	12 – 6	18 – 9	10 + 10	9 + 6

17 – 9

START

MAIL

10 + 10	8 + 9	6 + 5	8 + 8

Chapter Review

Name _____

Language and Math

Complete.
Use a word from the list.

Math Words

addends
difference
fact family
doubles
sum

Read these words.

1. 8 + 8 is a _____ fact.

2. 18 is the _____ of 9 + 9.

3. The _____ of 13 − 6 is 7.

4. 9 + 5 = 14 and 5 + 9 = 14 belong in the same

_____.

Concepts and Skills

Add or subtract.

5. 7 + 7 = ____   7 + 8 = ____   16 − 8 = ____

6. 7 + 5 = ____   17 − 9 = ____   6 + 8 = ____

7.

18	10	9	8	11	15	7
− 9	+ 10	+9	+ 8	− 6	− 8	+ 8

8.

7	8	3	6	9	2	5
3	8	7	4	9	7	5
+ 6	+ 2	+ 4	+ 5	+ 2	+ 9	+ 5

Add or subtract.

9. 10 − 10 = ____   9 + 8 = ____   18 − 9 = ____

10.

10	8	14	16	14	9	20
+ 0	+ 6	− 6	− 9	− 7	+ 9	−10

11.

15	20	18	17	7	6	8
− 9	+ 0	− 9	− 8	+ 7	+ 4	+ 8

12.

18	7	15	9	17	7	16
− 9	+ 8	− 6	+ 8	− 8	+ 9	− 9

Problem Solving

Write a number sentence to solve.

13. Jill has 5 stamps from Mexico and the same number of stamps from Spain. How many stamps does she have altogether?

____ + ____ = ____

____ stamps

14. Lou had 15 postcards from Chicago. He gave 8 of them away. How many does he have now?

____ − ____ = ____

____ postcards

Extra Practice

Name _____

$$\begin{array}{r} 6 \\ + 7 \\ \hline \end{array}$$

$$\begin{array}{r} 15 \\ - 8 \\ \hline \end{array}$$

$$\begin{array}{r} 9 \\ + 9 \\ \hline \end{array}$$

$$\begin{array}{r} 13 \\ - 8 \\ \hline \end{array}$$

$$\begin{array}{r} 7 \\ + 8 \\ \hline \end{array}$$

$$\begin{array}{r} 17 \\ - 9 \\ \hline \end{array}$$

$$\begin{array}{r} 6 \\ + 8 \\ \hline \end{array}$$

$$\begin{array}{r} 15 \\ - 9 \\ \hline \end{array}$$

$$\begin{array}{r} 9 \\ + 5 \\ \hline \end{array}$$

$$\begin{array}{r} 16 \\ - 7 \\ \hline \end{array}$$

$$\begin{array}{r} 7 \\ + 9 \\ \hline \end{array}$$

$$\begin{array}{r} 14 \\ - 7 \\ \hline \end{array}$$

$$\begin{array}{r} 18 \\ - 9 \\ \hline \end{array}$$

$$\begin{array}{r} 20 \\ - 10 \\ \hline \end{array}$$

$$\begin{array}{r} 9 \\ + 9 \\ \hline \end{array}$$

$$\begin{array}{r} 16 \\ - 8 \\ \hline \end{array}$$

Algebra & functions

Look at the numbers. What **patterns** do you see?

| 46 |
| 36 |
| 26 |
| 16 |

The rule is subtract **6**. What patterns do you see?

Subtract 6.

	− 6
46	40
36	30
26	20
16	10

Complete each table.

1.

	− 8
12	
13	
14	
15	

	+ 7
5	
6	
7	
8	

	− 9
14	
15	
16	
17	

2. Find the rule for each table. Then complete the table.

4	13
5	14
6	
7	

13	
14	7
15	
16	9

11	5
12	
13	7
14	

Name _____

Add.

1. 7 + 6 = ____   8 + 3 = ____   9 + 6 = ____

2. 8 + 8 = ____   8 + 9 = ____   8 + 7 = ____

3.
```
  8       3       7       6       5       9
  1       3       7       4       0       1
+ 9     + 9     + 2     + 2     + 6     + 6
```

Add or subtract.

4.
```
 16      18      15      15      15       9
- 9     - 9     - 9     - 8     - 7     + 9
```

```
 10      17       9      16       8      15
+10     - 9     + 8     - 8     + 7     - 6
```

Write a number sentence. Solve.

5. Julie has 9 postcards. She buys 5 more.
How many postcards does she have now?

_____ ◯ _____ = _____ postcards

Name _____

Stamps from Friends

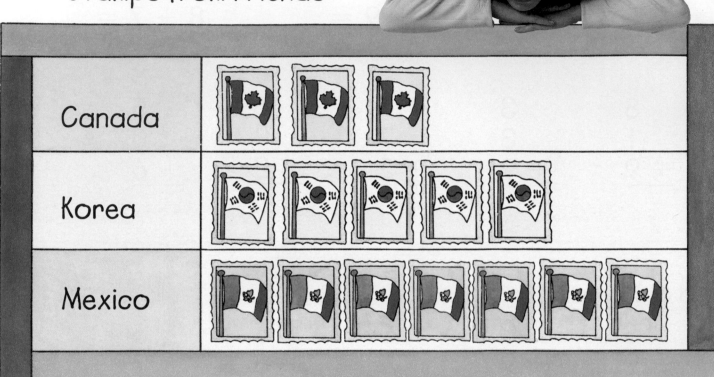

Canada	🇨🇦 🇨🇦 🇨🇦
Korea	🇰🇷 🇰🇷 🇰🇷 🇰🇷 🇰🇷
Mexico	🇲🇽 🇲🇽 🇲🇽 🇲🇽 🇲🇽 🇲🇽 🇲🇽

1. Write two addition sentences about the graph.

_____

_____

2. Write two subtraction sentences about the graph.

_____

_____

 You may want to put this page in your portfolio.

Name _____

Mathematical Reasoning

1. Don has 8 stamps. He just bought 6 more. How many stamps does he have now?

- ⬭ 16
- ⬭ 14
- ⬭ 12
- ⬭ 8

2. Max wrote 9 postcards. His sister wrote the same number of postcards. How many postcards did they write in all?

- ⬭ 9
- ⬭ 17
- ⬭ 18
- ⬭ 19

3. Lucy has 12 stamps. She used 3. How many stamps does Lucy have left?

- ⬭ 3
- ⬭ 7
- ⬭ 8
- ⬭ 9

Tell how you solved this problem.

_____

Number Sense

4. Which of these is a doubles fact?

- ⬭ 8 + 7
- ⬭ 8 + 8
- ⬭ 9 + 7
- ⬭ 7 + 9

5. Which names the number?

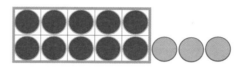

- ⬭ 10
- ⬭ 13
- ⬭ 15
- ⬭ 20

6. Which number sentence belongs in the same family of facts as 6 + 9 = 15?

- ⬭ 5 + 10 = 15
- ⬭ 9 − 6 = 3
- ⬭ 6 + 6 = 12
- ⬭ 15 − 6 = 9

Name _____

Algebra and Functions

7. Which number sentence can you use to solve this problem?

Al has 4 marbles. Pam has 3 marbles. How many marbles do Al and Pam have altogether?

- ⬭ $4 + 3 = 7$
- ⬭ $7 - 3 = 4$
- ⬭ $7 - 4 = 3$
- ⬭ $3 + 7 = 10$

8. Which sign goes in the box?

$3 \boxed{} 5 = 8$

- ⬭ $=$
- ⬭ $+$
- ⬭ $-$
- ⬭ $\times$

9. Julie has 7 books. She gives 3 away. Which shows how many books are left?

- ⬭ $3 + 7 = 11$
- ⬭ $7 - 5 = 2$
- ⬭ $7 - 3 = 4$
- ⬭ $11 + 7 = 14$

Measurement and Geometry

10. What time is shown on the clock?

- ⬭ 2:00
- ⬭ 2:30
- ⬭ 5:30
- ⬭ 6:00

11. Which figure is the same size as this?

- ⬭
- ⬭
- ⬭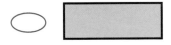
- ⬭

12. Which figure has 3 corners?

- ⬭
- ⬭
- ⬭
- ⬭

theme

Snap and Tap!

Use the Data

About how many children are in the parade?

What You Will Learn

In this chapter you will learn how to:

- Count, read, write, and show numbers to 100.

- Compare and order numbers to 100.

- Skip count to 100.

- Use logical reasoning to solve problems.

Snap and Tap!

Story by Marsha Comito
Illustrated by John Nez

Dear Family,

In Chapter 3 I will learn about tens and ones and numbers to 100. Here are new vocabulary words and an activity that we can do together.

Guess Again!

- Put about 60 beans in a bowl. Then have your child take a handful and put them on the table.

- Use tens numbers, such as 10, 20, and 30. Estimate about how many beans are in the pile.

- Then have your child count the beans.

- Ask your child if the estimate was more or less than the number he or she counted.

- Repeat the activity several times.

use

beans

pasta

Math Words

digit   56
↑ ↑
digit

ones   1**3**
↑
3 ones

tens   **2**5
↑
2 tens

ordinal numbers

Luke is third in line.
↓

1st 2nd 3rd 4th 5th

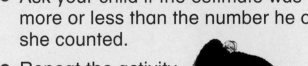

Additional activities at
www.mhschool.com/math

McGraw-Hill School Division

Learn

I can make a ten.

Math Words

tens
ones

__1__ ten = __10__ in all

Try it   Use to make tens.

1. 2 groups of ten

__2__ tens = __20__ in all

Workspace

2. 3 groups of ten

_____ tens = _____ in all

3. 4 groups of ten

_____ tens = _____ in all

 How many tens are there in 80?

Math at Home: Your child used cubes to make groups of 10.
Activity: Give your child 40 beans. Have your child put the beans into groups of 10.
Then have your child tell you how many groups of 10 there are.

Practice   Write how many.

4. 5 groups of ten   6 groups of ten

___5___ tens = __50__ in all   _____ tens = _____ in all

5. 7 groups of ten   8 groups of ten

_____ tens = _____ in all   _____ tens = _____ in all

6. 9 groups of ten   10 groups of ten

_____ tens = _____ in all   _____ tens = _____ in all

 Critical Thinking   Journal

7. Circle all the ways to show 28.

20 + 8   20 tens

2 tens 8 ones   28

---

8. Circle all the ways to show 53.

five ones   50 + 3

5 tens 3 ones   5 + 3

Learn

About how many musical instruments are in the picture?

You can estimate how many.

(about 10)   about 30

Try it

About how many notes are in each group?
Circle your estimate.

1.

about 10   about 30

2.

about 10   about 40

 **Sum it Up** How did you estimate?

 Math at Home: Your child estimated how many objects are in a group.
Activity: Put 10 beans in a group. Then have your child put a handful of beans in another group and estimate how many beans there are.

Practice About how many notes are in each group?
Circle your estimate.

3.

about 10   about 40

4.

about 20   about 50

5.

about 10   about 30

6.

about 20   about 50

Problem Solving

Number Sense

7. About how many shakers are in the picture? _____

8. Now count how many. _____

Name _____

Learn

You can regroup ones as tens and ones to show 23.

The digit 2 means 2 tens. The digit 3 means 3 ones.

23 ones

__2__ tens __3__ ones

tens	ones
2	3

Try it

Write how many tens and ones.
Use  and ▪.

1. 14 = __1__ tens __4__ ones

tens	ones
1	4

2. 25 = _____ tens _____ ones

tens	ones

3. 17 = _____ tens _____ ones

tens	ones

4. 30 = _____ tens _____ ones

tens	ones

Sum it Up
What are some different ways to show 56?

 Math at Home: Your child learned about tens and ones in 2-digit numbers.
Activity: Write some 2-digit numbers, like 53, and have your child tell you how many tens and how many ones.

eighty-three **83**

Practice Write how many tens and ones.

tens	ones
1	8

5. 18 = __1__ tens __8__ ones

tens	ones

6. 22 = _____ tens _____ ones

tens	ones

7. 87 = _____ tens _____ ones

tens	ones

8. 63 = _____ tens _____ ones

Write each number.

9. 5 tens 2 ones _____   8 tens 6 ones _____

10. 4 tens 7 ones _____   6 tens 3 ones _____

 Problem Solving

Draw a picture to solve.

11. Jan has 5 tens and 3 ones.
Merry has 2 tens and 3 ones.
How many more tens does
Jan have than Merry?

_____

Workspace

Learn

52 is fifty-two

These are word names for some numbers.

1	one	11	eleven	30	thirty
2	two	12	twelve	40	forty
3	three	13	thirteen	50	fifty
4	four	14	fourteen	60	sixty
5	five	15	fifteen	70	seventy
6	six	16	sixteen	80	eighty
7	seven	17	seventeen	90	ninety
8	eight	18	eighteen	100	one hundred
9	nine	19	nineteen		
10	ten	20	twenty		

Try it   Write each number.

1. seven __7__   fourteen _____   ninety _____

2. thirty _____   seventy-five _____   sixty-two _____

3. eighty-three _____   twenty-four _____   forty-nine _____

Write each number word.

4. 5 _____   16 _____   53 _____

5. 47 _____   99 _____   8 _____

Sum it Up   How do you write 73 as a word?

Math at Home: Your child learned to read and write numbers to 99 as words.
Activity: Give your child some 2-digit numbers, like 47, to write.

eighty-five **85**

Write each number.

6. eighty _80_   eleven _____   twenty _____

7. forty-two _____   sixty-three _____   forty-eight _____

8. fifty-six _____   thirty-one _____   seventy-nine _____

9. sixty-eight _____   eighty-one _____   ninety-five _____

Write each number word.

10. 9 _____   15 _____   46 _____

11. 40 _____   83 _____   92 _____

12. 39 _____   64 _____   76 _____

13. 26 _____   57 _____   28 _____

14. How are these numbers alike?

_____

15. How are they different?

_____

Learn

You can show a number in different ways.

5 tens and 3 ones = 53

50 + 3 = 53 expanded form

5 tens are 50.
The 5 means 50.
3 ones are 3.

Math Words

expanded form

Try it

Write each number in expanded form.

 28

1. $70 + 5$ _____   _____

  63

2. _____   _____   _____

 Sum it Up! How do you know if the 5 means 50 or 5 in the number 52?

 **Math at Home:** Your child learned the value of each digit in 2-digit numbers and wrote them in expanded form.
Activity: Say some 2-digit numbers and have your child tell the value of each digit.

eighty-seven **87**

Write each number in expanded form.

3. ___80 + 2___   _____   _____

4. _____   _____   _____

5. _____   _____   _____

6. _____   _____   _____

Algebra & functions

What could the missing number be? Explain the counting pattern.

7. | 23 | 33 | | 53 | 63 | 73 | _____

8. | 58 | | 78 | 88 | 98 | _____

Name _____

Draw Conclusions

Reading Skill   You can put together clues to help you draw conclusions.

Sara likes to dance. She invites friends to the dance party. Some friends dance in a circle. Others play in the band.

Solve.

1. Are more children dancing or playing in the band?

   _____

2. How many more? _____

3. Tell how you know. _____

4. Do you think Sara and her friends are having fun? Why?

   _____

 Math at Home: Your child read a story and used clues to draw conclusions.
Activity: Read your child a story. Then have your child draw some conclusions about the story.

Solve.

Sara and her friends like to line dance. They make up different steps. They cross their feet. They clap their hands.

5. How many rows of children are there? _____ rows

6. How many children are in each row? _____ children

7. How many children are there in all? _____ children

8. What if one more row of children joins them? How many children are dancing now? _____ children

9. Tell how you know. _____

10. Do you think Sara and her friends like to dance? Why?

_____

11. Why do you think Sara likes to dance?

_____

12. Use the picture. Write a problem. _____

Check Your Progress A

Name _____

About how many are in each picture?

1.

(about 10)   about 50   about 10   (about 30)

2. 6 groups of ten   4 groups of ten

_____ tens = _____ in all   _____ tens = _____ in all

Write how many tens and ones.

tens	ones

3. 56 = _____ tens   _____ ones

Write each number.

4. fifty-nine   twenty-one   forty-eight

_____   _____   _____

Write each number in expanded form.

5.   59   73

_____ + _____   _____ + _____

 Journal In the number 67, how do you know that the digit 7 means 7 and not 70?

Add.

1.

3	3	5	9	7	9	5
+ 3	+ 4	+ 9	+ 5	+ 8	+ 8	+ 7

Subtract.

2.

12	9	16	14	17	16	13
− 6	− 3	− 7	− 8	− 9	− 0	− 6

Add or subtract.

3. $9 + 2 =$ _____ $13 − 7 =$ _____ $5 + 8 =$ _____

4. $9 − 3 =$ _____ $12 − 6 =$ _____ $11 − 5 =$ _____

TECHNOLOGY LINK

Use Models to Show Expanded Form

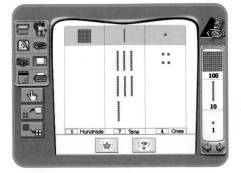

- Choose place value. Choose a mat.
- Stamp out 7 tens and 4 ones.
- What number is shown? _____
- Write the number in expanded form.

_____

1. Stamp out other numbers. Write in expanded form.

For more practice use Math Traveler™.

Name _____

Use Logical Reasoning

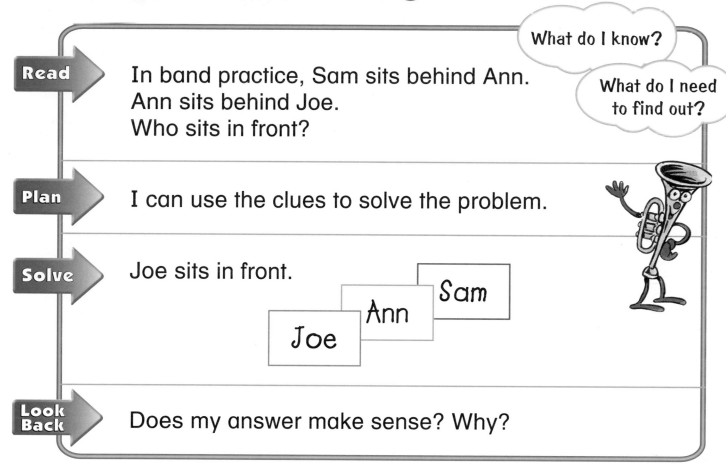

Read   In band practice, Sam sits behind Ann.
Ann sits behind Joe.
Who sits in front?

What do I know?

What do I need to find out?

Plan   I can use the clues to solve the problem.

Solve   Joe sits in front.

Sam

Ann

Joe

Look Back   Does my answer make sense? Why?

Use the clues. Solve.

1. In Sam's band, each group wears red, green, or blue.
   The trumpet players wear blue.
   The drum players don't wear red.
   What color do the drum players wear? _____

   The drum players wear _____ .

Explain how you used logical reasoning to help you solve the problems.

Math at Home: Your child solved problems by using logical reasoning.
Activity: Ask your child to describe how he or she solved problem 1.

Practice.

Read each problem. Then answer the questions.

Lin, Billy, and Ronny play in a band. They play a violin, a guitar, and a drum. Lin does not play violin. Billy does not play an instrument with strings.

2. What does Billy play? _____

3. What does Lin play? _____

4. What does Ronny play? _____

Todd, Walter, and Sam wear band uniforms. One uniform has 3 badges, one has 4 badges, and one has 5 badges. Sam's uniform has more than 4 badges. Todd's uniform has the fewest badges.

5. How many badges does Todd's uniform have? _____

6. How many badges does Walter's uniform have? _____

7. How many badges does Sam's uniform have? _____

8 Look at the picture. Write a problem about the picture.

Learn

You can compare numbers using <, >, and =.

Compare the tens first.

Math Words

compare
is less than <
is greater than >
is equal to =

Compare the tens.

Compare the ones.

22 is less than 25.

22 < 25

Compare the tens.

Compare the ones.

25 is greater than 22.

25 > 22

Compare the tens.

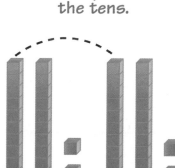

Compare the ones.

22 is equal to 22.

22 = 22

Try it

Compare. Use <, >, or =. You can use and .

1. 42 53   84 77   17 17   28 32

2. 58 61   42  42   92 87   33 29

Sum it Up

How can you show that one number is greater than or is less than another number?

Math at Home: Your child learned how to compare numbers.
Activity: Ask your child to tell three numbers greater than 25 and three numbers less than 80.

Practice

Compare. Use <, >, or =.
Use and if you like.

1. 90 ⊂<⊃ 91   82 ◯ 76   21 ◯ 21   43 ◯ 36

2. 40 ◯ 40   56 ◯ 58   79 ◯ 71   67 ◯ 67

3. 36 ◯ 51   72 ◯ 72   47 ◯ 35   53 ◯ 9

4. 39 ◯ 39   6 ◯ 68   7 ◯ 25   26 ◯ 22

5. 34 ◯ 4   40 ◯ 35   19 ◯ 33   71 ◯ 71

6. 95 ◯ 99   73 ◯ 8   49 ◯ 49   42 ◯ 51

7. 80 ◯ 78   46 ◯ 51   76 ◯ 76   68 ◯ 67

8. I am a number.
I am greater than
4 tens and 3 ones
and less than
4 tens and 5 ones.
What number am I? _____

9. I am a number.
I am less than
7 tens and 7 ones
and greater than
7 tens and 5 ones.
What number am I? _____

Learn

You can write numbers in order.

1	2	3	4	5	6	7	8	9	10
11	12	13	14	15	16	17	18	19	20
21	22	23	24	25	26	27	28	29	30
31	32	33	34	35	36	37	38	39	40
41	42	43	44	45	(46)	47	[48]	49	50
51	52	53	54	55	56	57	58	59	60
61	62	63	64	65	66	67	68	69	70
71	72	73	74	75	76	77	78	79	80
81	82	83	84	85	86	87	88	89	90
91	92	93	94	95	96	97	98	99	100

Math Words

order
before
after
between

46 comes before 47.
48 comes after 47.
47 comes between 46 and 48.

Try it

Write the number that comes just before.

1. 22 | 23   ___ | 54   ___ | 73   ___ | 50

Write the number that comes just after.

2. 31 | ___   46 | ___   78 | ___   29 | ___

Write the number that comes between.

3. 61 | ___ | 63   28 | ___ | 30   49 | ___ | 51   96 | ___ | 98

 **Sum it Up** What is the number just before 36? What is the number just after 36?

 Math at Home: Your child ordered numbers that come before, after, and between other numbers.
Activity: Pick a number from 10 to 99. Have your child tell you the number that comes before and after the number.

You can use the hundreds chart.

Write the number that comes just before.

4.
| | 26 | | | 14 | | | 56 | | | 32 |

5.
| | 79 | | | 90 | | | 41 | | | 63 |

Write the number that comes just after.

6.
| 28 | 27 | | 34 | | | 42 | | | 50 | |

7.
| 61 | | | 86 | | | 99 | | | 13 | |

Write the number that comes between.

8.
| 14 | | 16 | | 53 | | 55 | | 25 | | 27 | | 80 | | 82 |

9.
| 48 | | 50 | | 59 | | 61 | | 32 | | 34 | | 61 | | 63 |

Number Sense

10. Which number is closest to 50?
How do you know?
Circle your answer.

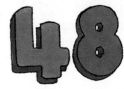  

Learn

You can skip count by twos.
This is a counting pattern.

 bells

2   4   6   8   10   12

Try it

Skip count the holes by threes. Write the number.

1.

____ ____ ____ ____ ____ ____

Skip count the bells by fives. Write the number.

2.

____ ____ ____ ____ ____ ____

Sum it Up Explain how to skip count by twos.

 Math at Home: Your child counted by twos, threes, fours, and fives.
Activity: Have your child count aloud by twos, threes, fours, and fives.

1	2	3	4	5	6	7	8	9	10
11	12	13	14	15	16	17	18	19	20
21	22	23	24	25	26	27	28	29	30
31	32	33	34	35	36	37	38	39	40
41	42	43	44	45	46	47	48	49	50
51	52	53	54	55	56	57	58	59	60
61	62	63	64	65	66	67	68	69	70
71	72	73	74	75	76	77	78	79	80
81	82	83	84	85	86	87	88	89	90
91	92	93	94	95	96	97	98	99	100

4. Skip count by twos. Color each number yellow.

5. Skip count by threes. Color each number green.

6. Skip count by fours. Color each number red.

7. Skip count by fives. Color each number blue.

Algebra & functions

8. Look at the hundreds chart.
Find a counting pattern.
Write it. _____

_____

Name _____ **Odd and Even Numbers**

Learn

Math Words

even

odd

I cannot break 5 cubes into 2 equal parts. 5 is an odd number.

I can break 6 cubes into 2 equal parts. 6 is an even number.

odd   even

Try it

Write if the number is even or odd.
You can use  to help.

1. 10 _even_   11 _____   15 _____

2. 18 _____   19 _____   22 _____

3. 17 _____   16 _____   75 _____

Sum it Up

How can you tell if a number is even or odd?

McGraw-Hill School Division

🏠 **Math at Home:** Your child identified even and odd numbers.
Activity: Tell your child a number between 1 and 100.
Have your child tell you if the number is even or odd.

one hundred one **101**

Write if the number is even or odd.

4. 23 _odd_   26 _____   86 _____

5. 37 _____   42 _____   99 _____

Write the next three even numbers.

6. 12 ___ ___ ___   24 ___ ___ ___

7. 76 ___ ___ ___   36 ___ ___ ___

8. 28 ___ ___ ___   48 ___ ___ ___

Write the next three odd numbers.

9. 51 ___ ___ ___   25 ___ ___ ___

10. 67 ___ ___ ___   33 ___ ___ ___

11. 49 ___ ___ ___   91 ___ ___ ___

Spiral Review and Test Prep

Choose the correct answer. Fill-in the ◯.

12. Which number is odd?

- ◯ 22
- ◯ 23
- ◯ 24
- ◯ 26

13. What is the turnaround fact for $9 + 8 = 17$?

- ◯ $1 + 7 = 8$
- ◯ $8 + 9 = 17$
- ◯ $9 - 8 = 1$
- ◯ $9 + 8 = 17$

Learn

Math Words

ordinal number

Ordinal numbers tell about position.

1st	2nd	3rd	4th	5th	6th	7th	8th	9th	10th
first	second	third	fourth	fifth	sixth	seventh	eighth	ninth	tenth

Try it   Circle the correct position.

1.

third

(fourth)

fifth

2.

first

third

seventh

3.

ninth

second

fourth

4.

second

third

sixth

 **Sum it Up**   Draw a picture to show the third person in a line.

Math at Home: Your child used ordinal numbers from first through tenth.
Activity: Have your child put four objects in a line and tell you which is first, second, third, and fourth.

one hundred three   **103**

Follow each direction.
Start at the left each time.

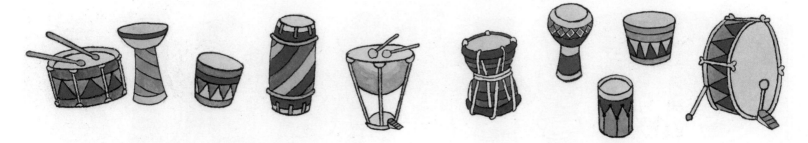

5. Circle the third drum in green.

6. Circle the ninth drum in red.

7. Circle the fifth drum in blue.

8. Circle the eighth drum in orange.

9. How many drums are in front of the sixth drum? _____

10. How many drums are in front of the second drum? _____

Answer each question.

11. There are 12 children in the line dance. How many are in front of the eighth child? _____

12. There are 12 children in the line dance. How many are behind the ninth child? _____

Logical Reasoning

13. A trumpet is behind a flute and in front of a drum on a store shelf. Tell the order. _____

Name _____

Decorate a Drum

You want to decorate a drum.
Here are the parts you can buy.
You have 50¢.

Plan which parts you want to buy.

Make a list.

Workspace

1. _____

_____

_____

_____

2. How much did you spend?
Show how you found out.

_____

_____

3. Draw a picture of your drum.
Then color it.

4. **What if** you buy one more item?
How much more will your
drum cost?

_____

_____

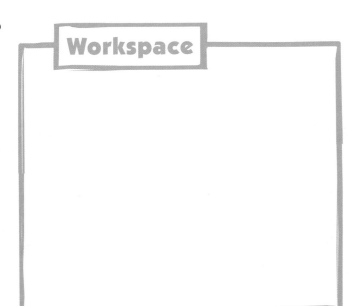

Workspace

Name _____

What Sounds Can You Hear?

Musical shakers can make many sounds.

What to do

1. Cut each tube a different length. Then fill with two spoonfuls of beans.

2. Tape the end of each tube.

3. Shake each tube. Listen to the sounds.

4. Shake each tube again. Describe the sounds you hear.

You Will Need
paper tubes
dried beans
tape
spoon
scissors

McGraw-Hill School Division

MR1.1

What did you find out?

1. Are the sounds alike or different?
Explain your answer.

_____

_____

2. Which tube sound is the loudest?
Which tube sound is the softest?

_____

_____

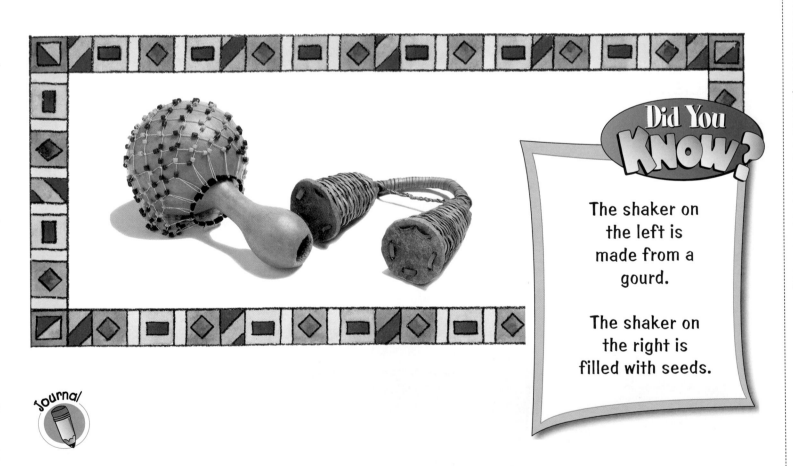

Did You KNOW?

The shaker on the left is made from a gourd.

The shaker on the right is filled with seeds.

Journal

Want to do more?

3. Use 2 tubes that are the same length. Put 20 beans in one tube and 60 beans in the other tube. Shake both tubes. How does the number of beans affect the sound?

Check Your Progress B

Name _____

Compare. Use <, >, or =.

1. 37 ◯ 52   49 ◯ 41   70 ◯ 44

2. 16 ◯ 61   28 ◯ 28   56 ◯ 60

Write each missing number.

3. | 19 | | 21 | | 42 | 43 | | | 98 | 99 | |

4. | 66 | 67 | | | | 56 | 57 | | 28 | | 30 |

Skip count by fives.

5. 5   1 0   _____   _____   _____   _____   _____   _____

Tell if each number is even or odd.

6. 19 _____   24 _____   30 _____   65 _____

7. 13 _____   51 _____   96 _____   72 _____

Answer each question.

8. Tom is older than Rob. Rob is not older than Sam. Sam is younger than Tom. Who is oldest? _____

9. There are 10 people in line. How many people are behind the ninth person? _____

Name_____

What's My Number?

- Play with a partner. Take turns.

- Do not let your partner see your hundreds chart.

- Circle one number. Give a clue, such as, "My number is less than 70."

- Your partner asks questions to guess your number.

- Answer yes or no until your number is named.

1	2	3	4	5	6	7	8	9	10
11	12	13	14	15	16	17	18	19	20
21	22	23	24	25	26	27	28	29	30
31	32	33	34	35	36	37	38	39	40
41	42	43	44	45	46	47	48	49	50
51	52	53	54	55	56	57	58	59	60
61	62	63	64	65	66	67	68	69	70
71	72	73	74	75	76	77	78	79	80
81	82	83	84	85	86	87	88	89	90
91	92	93	94	95	96	97	98	99	100

Name _____

Language and Math

Complete. Use a word from the list.

1. 6 comes just _____ 7.

2. 4, 8, and 10 are _____ numbers.

3. In the number 35, the 3 is in the _____ place.

Concepts and Skills

Write how many.

4. 8 groups of 10 _____ tens = _____ in all

5. 3 groups of 10 _____ tens = _____ in all

6. 1 group of 10 _____ ten = _____ in all

7. 7 groups of 10 _____ tens = _____ in all

8. There are 10 people in line. How many people are in front of the third person?

9. There are 8 people in line. How many people are behind the fourth person?

_____

McGraw-Hill School Division

Write how many tens and ones.

8.

26 = _____ tens _____ ones

tens	ones

9.

43 = _____ tens _____ ones

tens	ones

10.

71 = _____ tens _____ ones

tens	ones

Compare. Use <, >, or =.

11. 23 ◯ 19   44 ◯ 50   65 ◯ 65

Write the number that comes just before.

12. _____ | 43   _____ | 50   _____ | 91

Skip count by fives.

13. 15   20   _____ _____ _____ _____ _____

Problem Solving

14. Ann's tuba is quieter than Cindy's trumpet. Charlie's trombone is louder than Ann's tuba. Cindy's trumpet is louder than Charlie's trombone. Which instrument is the loudest? _____

Name _____

Color.

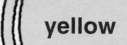

blue

odd numbers

yellow

even numbers

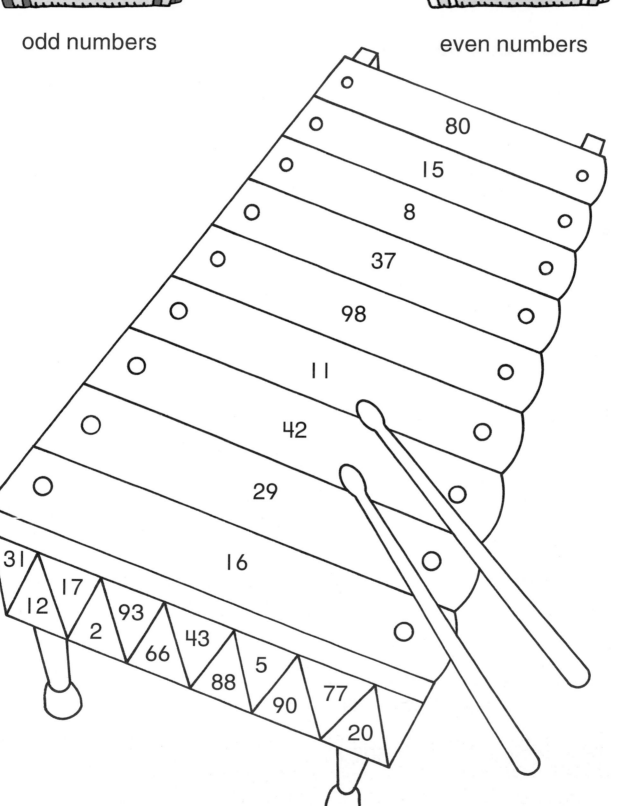

Is 23 closer to 20 or 30?
Use the number line to find out.

Find 23. 23 is between 20 and 30.
20 is the nearest ten.

Use the number lines.

1.

Is 18 closer to 10 or 20? How do you know?

closer to $\underline{20}$ _____

2.

Is 24 closer to 20 or 30? How do you know?

closer to _____ _____

3.

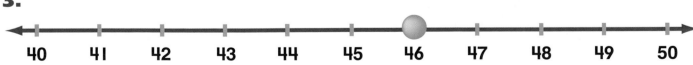

Is 46 closer to 40 or 50? How do you know?

closer to _____ _____

Name _____

Write how many tens and ones.

1. 17 = _____ tens _____ ones

2. 45 = _____ tens _____ ones

Write each number or number word.

3. sixteen _____   seventy-four _____

4. forty-one _____   eighty-six _____

5. 9 _____   35 _____

6. 51 _____   100 _____

Write each missing number.

7. | 23 | | 25 | | 67 | 68 | | | 18 | 19 | | | | 59 | 60 |

Compare. Use <, >, or =.

8. 33 ◯ 29   54 ◯ 45   19 ◯ 63   78 ◯ 68

Skip count by fours.

9. 8   12   _____   _____   _____   _____   _____

Solve.

10. The trumpet is bigger than the flute.
The trombone is bigger than the trumpet.
Which instrument is biggest?   _____

I'm thinking of a number.

Choose a number.

Your number _____

How many tens are in your number? _____

How many ones are in your number? _____

Is your number even or odd? _____

What number comes after your number? _____

What number comes before your number? _____

Name _____

Choose the correct answer.

Statistics, Data Analysis, and Probability

1. George made a tally chart showing the weather. Which kind of weather happened for 6 days?

○ ☀
○ ☁
○ 🌧
○ 💨

2. Which doesn't belong?

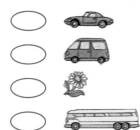

○ 🚗
○ 🚐
○ 🌸
○ 🚌

3. Look at the number pattern. What could the next number be?

8   10   12   14   _____

○ 6
○ 12
○ 14
○ 16

Mathematical Reasoning

4. Lois plays 3 games. Pam plays 5 games. How many games do they play altogether?

○ 2 games
○ 3 games
○ 5 games
○ 8 games

5. Ling has 11 crayons. She gives away 7 of them. How many does she have left?

○ 4
○ 7
○ 11
○ 18

How did you solve this problem?

_____

6. Jenny has 15 pennies. She got 3 more. How many pennies does she have now?

○ 15
○ 17
○ 18
○ 20

Spiral Review and Test Prep

Number Sense

7.
$$\begin{array}{r} 8 \\ + 9 \\ \hline \end{array}$$

- ○ 15
- ○ 16
- ○ 17
- ○ 18

8. Which are all odd numbers?

- ○ 3, 4, 7
- ○ 11, 13, 19
- ○ 16, 17, 18
- ○ 4, 8, 10

9. Which shows 47?

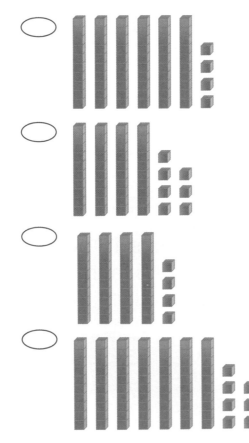

Algebra and Functions

10. Which number sentence can you use to solve this problem?

Cindy has 7 cards. Carlos has 3 cards. How many more cards does Cindy have?

- ○ $4 + 3 = 7$
- ○ $7 - 3 = 4$
- ○ $7 - 4 = 3$
- ○ $3 + 7 = 10$

11. What goes in the box?

$9 \boxed{} 5 = 4$

- ○ $=$
- ○ $+$
- ○ $-$
- ○ $\times$

12. What is the favorite bird?

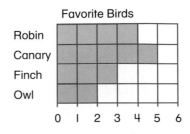

- ○ Robin
- ○ Canary
- ○ Finch
- ○ Owl

BUMPER CARS 40¢

GO FISH

RING OO TOSS

55¢

20¢

RIDDLE HOUSE

SCOOP the DUCK 75¢

theme
At the Fair

Use the Data

Look at the picture. How many games cost more than 50¢?

What You Will Learn

In this chapter you will learn how to:

- Identify coins, bills, and their value.
- Count to find how much money.
- Compare money amounts.
- Use coins to solve problems.

Let's Go to the Fair!

Story by Sharon Montoya
Illustrated by Margeaux Lucas

MATH AT HOME

Dear Family,

In Chapter 4, I will learn about money. Here are new vocabulary words and an activity that we can do together.

What's the Price?

Math Words

penny

1¢
one cent

nickel

5¢
five cents

dime

10¢
ten cents

quarter

25¢
twenty-five cents

dollar

$1.00
one dollar

- Put some price tags below 50¢ on some household objects.

- Place objects with price tags on a table. Ask your child to use coins to show the price.

- Have your child tell you which object has the lowest price and which object has the highest price.

use

small household objects

price tags

Additional activities at
www.mhschool.com/math

Learn

I penny,
I cent, I¢

I nickel,
5 cents, 5¢

I dime,
10 cents, 10¢

I quarter,
25 cents, 25¢

Math Words

penny
nickel
dime
quarter
cent ¢

Try it

Use coins to show each price. Draw the coins.

1.

8¢

Workspace

2.

23¢

Workspace

Sum it Up

How many cents are in a penny, in a nickel, in a dime, and in a quarter?

Math at Home: Your child explored how to use coins to make money amounts up to 25¢.
Activity: Give your child a price up to 25¢. Ask your child to show coins for that amount.

Use coins.
Show each price in two ways.
Draw the coins.

3. 7¢

4. 15¢

5. 18¢

6. 26¢

 Problem Solving

 Algebra & functions

Tina put 12¢ in the bank.
Jordan put the rest of the coins
in the bank. How much money
did Jordan put in the bank? _____

Learn

Find the total amount.
Start with the coins that have
the greatest value.

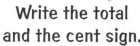
Write the total
and the cent sign.

10¢   20¢   30¢   40¢   41¢   42¢   42¢ total

Try it

Count to find the total amount.

1.

5¢   10¢   15¢   20¢   25¢

25¢
total

2.

_____ _____ _____ _____ _____

total

3.

_____ _____ _____ _____ _____ _____

total

4.

_____ _____ _____ _____ _____ _____

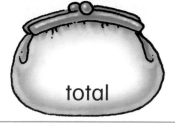
total

Sum it Up

How can you order coins to make them easier to count?

Math at Home: Your child counted groups of coins to 99¢.
Activity: Give your child a group of coins to count. Help your child
arrange the coins in order starting with the greatest value.

Count to find the total amount.
Do you have enough money to buy the toy?
Circle yes or no.

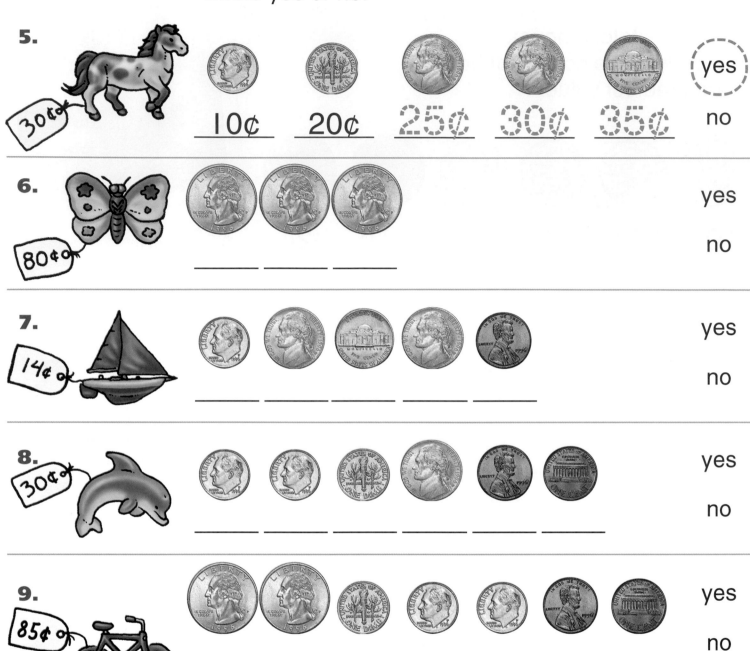

5. 30¢

10¢   20¢   25¢   30¢   35¢

(yes)

no

6. 80¢

_____   _____

yes

no

7. 14¢

_____   _____   _____

yes

no

8. 30¢

_____   _____   _____

yes

no

9. 85¢

_____   _____   _____   _____

yes

no

 Problem Solving

Workspace

Draw a Picture

11. Josie has some nickels and dimes. She
buys a 20¢ ticket to play toss the ring.
How many different ways can Josie
pay for her ticket? Show the ways. _____

Learn

I can use 2 quarters to pay.

I can use one half dollar to pay.

Sack Race 50¢

Math Words

half dollar

25¢   25¢

half dollar 50¢

Try it   Count.
Match each group of coins to a price.

1.

70¢

75¢

2.

50¢

55¢

   How would you show 50¢ using two coins? Explain.

    Math at Home: Your child counted groups of coins that included half dollars.
Activity: Ask your child to show you how to count a group of coins.

McGraw-Hill School Division

Practice
Count each group of coins.
Write the total amount.

2.  67¢

3.  _____

4.   _____

5.   _____

Circle the coins for each amount.

6.  45¢

7.  83¢

Spiral Review and Test Prep

Choose the correct answer.

8. How much money is shown?

- ○ 46¢
- ○ 61¢
- ○ 71¢
- ○ 81¢

9. Which of these is a doubles fact?

- ○ 8 + 3
- ○ 9 + 4
- ○ 9 + 9
- ○ 7 + 8

Learn

Use the fewest number of coins to pay.

> Start with the coin that has the greatest value.

> I can use 1 quarter, 1 dime and 1 nickel to make 40¢.

LEMONADE 40¢

Try it

Use the fewest number of coins to pay.
Circle the coins.

1.

2.

3.

Sum it Up

How does starting with the coin with the greatest value help you find the fewest coins?

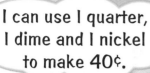 **Math at Home:** Your child learned how to show money amounts using the fewest coins.
Activity: Have your child show you how to make 32¢ using the fewest coins.

one hundred twenty-seven **127**

Practice

Use the fewest number of coins to pay.
Circle the coins.

4.  35¢

5.  45¢

6.  85¢

7.  95¢

Critical Thinking

8. Which group of coins does not belong? Explain.

Name _____

Learn

Darcy buys a ticket for 48¢.
She gives the clerk 50¢.
How much change does Darcy get back?

Count up to 50¢.
Start at 48¢ and say 49¢, 50¢.

2¢ ___ change

So, 2¢ is the change.

Try it   Complete the table.

	Ticket price	You pay	Count up to find your change
1.	Pony Ride 22¢ 22¢	(quarter)	3¢ ___ change
2.	Bumper Car Ride 37¢ 37¢	(quarter, nickel, dime)	_____ change

 How can you count up to make change? Give an example.

 Math at Home: Your child learned how to count up from a price to make change.
Activity: Give your child a price less than a dollar and an amount given as payment and have your child count up to make change.

Practice   Complete the table.

	Ticket price	You pay	Count up to find your change
3.	TOSS the RING 24¢ 24¢	quarter	_ _ ¢ change
4.	SCOOP THE DUCK 16¢ 16¢	dime, dime	_____ change
5.	Merry-Go-Round 32¢ 32¢	quarter, dime	_____ change
6.	Go Fishing 53¢ 53¢	half dollar, nickel	_____ change

 Problem Solving

7. Which set of coins has the greater value? Explain.

_____

Name _____

Cause and Effect

Reading Skill   You can read to find out why something happens to help you solve problems.

Jo wants to go to the Fair!
Jo does jobs for her
mother to make money.
She feeds the cats.
She walks the dog.
Her mother gives Jo 4
dimes, 2 nickels, and
5 pennies.

Solve.

1. What jobs does Jo do?

_____

2. Why does Jo do jobs for her mother?

_____

3. What does she get for doing the jobs?

_____

4. How much money does Jo have?

_____

Math at Home: Your child learned why an event happened in a story and what happened as a result of these events. **Activity:** Have your child tell about a new activity they learned how to do. What happened? Why did it happen?

Solve.

Sam wants to play games at the Fair. Sam helps his grandpa make lemonade. His grandpa gives him 3 quarters. On Saturday Sam goes to the Fair to play games.

5. How does Sam help his grandpa?

_____

6. Why does Sam help his grandpa?

_____

7. What does Sam get for helping? _____

8. How much money does Sam get? _____

9. Use the story.
   Write a problem.

_____

Check Your Progress A

Name _____

Use coins.
Show each price in two ways.
Draw the coins.

1.

Count to find the total amount.
Do you have enough money to buy the item?
Circle yes or no.

2.

yes

no

_____   _____   _____

Count each group of coins. Write the total amount.

3.   _____

Use the fewest coins to pay. Circle the coins.

4.

Complete the table.

Ticket price	You pay	Count up to find your change
5. Hopping Contest 17¢ 17¢	(dime) (dime)	_____ change

Add or subtract.

1.
$$\begin{array}{r} 4 \\ +4 \\ \hline \end{array} \qquad \begin{array}{r} 4 \\ +5 \\ \hline \end{array} \qquad \begin{array}{r} 9 \\ -4 \\ \hline \end{array} \qquad \begin{array}{r} 13 \\ -6 \\ \hline \end{array} \qquad \begin{array}{r} 7 \\ +7 \\ \hline \end{array} \qquad \begin{array}{r} 15 \\ -9 \\ \hline \end{array} \qquad \begin{array}{r} 14 \\ -8 \\ \hline \end{array}$$

Write the number.

2. fifty-six _____   thirty _____   sixty-five _____

Write the next three even numbers.

3. 14   16   _____   _____   _____

TECHNOLOGY LINK

Count Money
- Use money.
- Choose a mat to show one amount.
- Stamp out 6 quarters, 4 nickels, and 5 pennies.
- What is the amount? _____
- Stamp out other coins. Tell the amount.

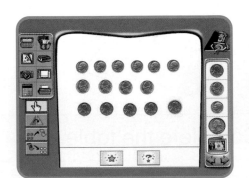

For more practice use Math Traveler.™

What do I know?

What do I need to find out?

Name _____

Act It Out

Read ▶ Phil has dimes and nickels in his pocket. In how many ways can he pay for a 35¢ ticket to the fun house?

Plan ▶ I can use coins and act it out to solve this problem.

Solve ▶

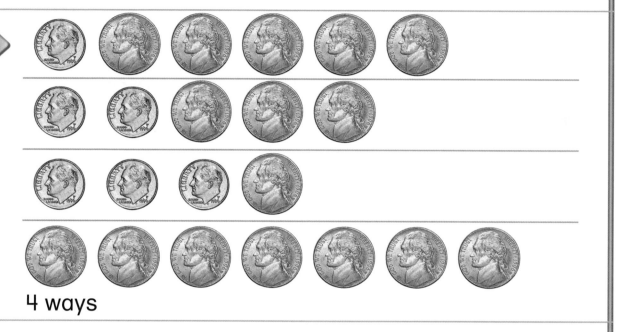

4 ways

Look Back ▶ Does my answer make sense? Why?

Work with a partner.
Use coins. Solve.

1. Jill has quarters, dimes, and nickels in her purse. In how many ways can she buy a 30¢ bottle of water.

_____ ways

2. How many nickels does it take to make 55¢?

_____ nickels

Math at Home: Your child used the strategy Act It Out to solve problems.
Activity: Give your child some coins and an item priced less than a dollar. Have your child use the coins to count your change for the item.

one hundred thirty-five

McGraw-Hill School Division

Use coins.
Work with a partner. Act it out.

3. Darren has quarters, dimes, and nickels in a bag. In how many ways can he buy a ticket for 25¢?

_____ ways

4. Sally has quarters, dimes, and nickels. In how many ways can she buy a ticket for the airplane ride that costs 40¢?

_____ ways

5. Alex has 5 dimes. He bought a ticket for 45¢. How much change should he get back?

_____

6. Lenny has a half dollar and a quarter. He bought a chili dog for 65¢. How much change should he receive?

_____

7. How many dimes make 80¢?

_____ dimes

8. How many nickels make 75¢?

_____ nickels

9. Look at the picture. Use money amounts. Write a problem.

_____

_____

RIDDLE HOUSE 50¢

RIDDLE HOUSE

Learn

Lynn and Max have money to spend at the Fair.
How much does each have?

Write the dollar sign ($) first.
Then write the decimal point.
to separate dollars
from cents.

| $1.00 | $1.25 | $1.35 | $1.36 |
Lynn has $1.36.

$0.25 $0.50 $0.75 $1.00 $1.05
Max has $1.05.

Try it

Count the money. Write the total amount.

1.

$1.00 $1.25 $1.50
Total

2.

_____ _____ _____ _____ _____ $ _____
Total

Sum it Up

Does it take more quarters or dimes to make $1.00.
Explain.

Math at Home: Your child counted dollars and cents.
Activity: Give your child one dollar and some coins and have your child count the total
amount of money.

Practice

Count the money.
Write the total amount.

Remember $0.51 is the same as 51¢.

3.

$0.25 $0.50 $0.75 $0.80 $0.81 $0.82 $0.83 $0.84   **$0.84**

Total

4.

_____ _____ _____ _____ _____ $_____

Total

5.

_____ _____ _____ _____ _____ $_____

Total

6.

_____ _____ _____ _____ $_____

Total

 ## Spiral Review and Test Prep

Choose the correct answer.

7. Jake bought an ear of corn
for 30¢. He gave the clerk fifty
cents. How much change did
Jake get?

◯ 15¢
◯ 20¢
◯ 25¢
◯ 30¢

8. 8 bugs are on a leaf.
3 fly away.
How many are left?

◯ 6
◯ 5
◯ 4
◯ 3

Name _____   **Compare Money**

Learn

Jen has this money in her pocket.

$1.00 $1.25 $1.50 $1.60 $1.65

$1.65 each

Does Jen have enough money
to buy a cap?

(Yes)   No

Try it

Count. Is there enough money to buy each item?
Circle yes or no.

1.

$1.35

$1.00 $1.25 $1.35 $1.45

yes

no

2.

$1.79

_____   _____   _____

yes

no

Sum it Up

How do you compare money amounts?

 Math at Home: Your child compared amounts of money.
Activity: Give your child two amounts of money and have your child tell which amount
is more.

one hundred thirty-nine **139**

McGraw-Hill School Division

Count. Is there enough money to buy each item? Circle yes or no.

3.

yes

(no)

$1.00 $1.25 $1.50 $1.60

4.

yes

no

_____ _____ _____ _____

5.

yes

no

_____ _____ _____ _____ _____

6.

yes

no

_____ _____ _____

💡 **Problem Solving**

Estimation

7. Mike has 2 quarters and 20 pennies. Does he have more than $1.00 or less than $1.00? _____

Name _____

Plan a Fair Day!

You want to go to the Fair.
Here is what you can do.
You have 99¢.
Buy some tickets.

TICKETS

BUMPER CAR RIDES 40¢

GO FISH 55¢

SCOOP the DUCK 25¢

TOSS the RING 20¢

LEMONADE 40¢

POPCORN 30¢

_____

You Decide!

1. Decide which tickets you want to buy. Make a list. Draw your tickets.

_____

_____

_____

_____

_____

Workspace

2. How did you decide which tickets to buy?

_____

_____

Workspace

3. How much money did you spend?

_____

_____

4. What if you had $2.00. Which tickets would you buy? Explain.

Workspace

_____

_____

_____

_____

Name _____

How Does a Seesaw Work?

A seesaw is a lever.
A lever is a simple machine.

What to do

1. Tape the pencil to a table.

2. Place the ruler over the pencil.

3. Put 1 penny on one end of ruler.
   Put 4 pennies on the other end.
   What happens?

4. Now remove the tape.
   Put 1 penny at one end of ruler.
   Put 4 pennies at the other end.
   Move the pencil to balance the ruler.
   What do you notice?

Other Levers

Results with Pencil Taped

Results with Pencil not Taped

A baseball bat helps a player hit the ball fast.

A fishing pole helps lift a fish out of the water.

What did you find out?

1. What happened to the ruler the first time? _____

2. What happened the second time? _____

3. Did the ruler balance when you moved the pencil? _____

4. Put 5 pennies on each end of the ruler.

Does the ruler balance? Explain. _____

_____

 Want to do more?

Use coins to show 10 cents two ways and put on the ends of the ruler. Predict what you think will happen.

Then try it. _____

Check Your Progress B

Name _____

Count the money.
Write the total amount.

1.

_____   _____

Count. Is there enough money to buy each item?
Circle yes or no.

2.     yes

no

3.     yes

no

Use coins. Solve.

4. Betsy has a jar of quarters, dimes and nickels.
In how many ways can she buy a ticket for 35¢?

_____

Name_____

Fun at the Fair

FINISH

Sell Your PARTNER a TOY PIG for 5¢

Buy a TOY CAT from Your PARTNER

Put 25¢ in Your PARTNER'S BANK

Give Your Partner 25¢ for a RING TOSS TICKET

- Put your marker on **Start**.

- Take turns. Flip a coin.

- Heads- you move one space.

- Tails- you move two spaces.

- Follow the directions in the space.

- Play until both players reach **Finish**.

- The player with the most money wins.

Your PARTNER puts 10¢ in Your BANK

Put 5¢ in Your PARTNER'S BANK

Your PARTNER Gives you 25¢ for a GO FISH TICKET

PUT 25¢ in Your PARTNER'S BANK

Sell a PONY TOY to Your PARTNER for 25¢

START

Your PARTNER Gives you 10¢ for a Hot DOG

Give Your PARTNER 5¢ to SCOOP a DUCK

Take 25¢ from Your PARTNER'S BANK

Name _____

Language and Math

Complete.
Use a word from the list.

Read these words.

Math Words

dime
half dollar
nickel
one dollar
quarter

1. A _____ is the same as a 10 pennies.

2. A _____ is the same as a 25 pennies.

3. A _____ is the same as a 5 pennies.

4. A _____ is the same as 50¢.

Concepts and Skills

Match.

5.

quarter   dime   nickel   penny

Count on to find the total amount.

6.

_____   _____   _____   _____ Total

_____   _____   _____   _____ Total

Count the money. Write the total amount.

7.

_____ _____ _____ _____

Complete the table.

Ticket price	You pay.	Count up to find your change.
8. Ride the Waves 22¢ 22¢	(quarter)	_____ change
9. Water Slide 33¢ 33¢	(quarter)(dime)	_____ change

Problem Solving

Use coins. Solve.

10. David has dimes, and nickels in a bag. How many ways can he buy a ticket for 20¢?

Name _____

Count. Write how much money.

1.

_____

2.

_____

3.

_____

4.

_____

5.

_____

Carmen has 7 coins that total $1.00.
She has two quarters. Draw the other coins.

$0.25 $0.50   _____   _____   _____   _____   _____

Carmen has 2 quarters and 5 dimes.

Each child puts $1.00 in coins into the vending machine.
Write the number of each coin that each person used.

CHILD		QUARTER	DIME	NICKEL
LINDA	8 coins	2		
STEVE	9 coins	2		
KYLE	10 coins	2		
DONNA	11 coins	2		
BEN	12 coins	2		

Carmen sees patterns.
What is the pattern for the total
number of coins used by each child?

_____

What patterns do you see for the number
of dimes and nickels used by each child?

_____

Name _____

Draw lines to match.

1.

one dollar   quarter   dime   half dollar

Count to find the total amount.

2.

_____   _____   _____   _____   _____   _____   total

Count. Is there enough money to buy each item? Circle yes or no.

3.

$2.42

yes

no

4.

$3.35

yes

no

Find the change.

| Ticket price | You pay. | Count up to find your change. |

5.

MOONWALK 35¢   35¢

_____ change

Use money if you want to.

Show different numbers of coins and bills to make each amount two ways.

1. 35¢				
2. 60¢				
3. $1.00				
4. $2.20				

 **Portfolio**

You may want to put this page in your portfolio.

Name _____

Measurement and Geometry

1. About how many paper clips wide is the piece of paper?

- ⚪ 2
- ⚪ 3
- ⚪ 6
- ⚪ 8

2. What is the date of the second Tuesday in May?

M	A	Y				
S	M	T	W	Th	F	S
				1	2	3	4
5	6	7	8	9	10	11
12	13	14	15	16	17	18
19	20	21	22	23	24	25
26	27	28	29	30	31	

- ⚪ May 7
- ⚪ May 13
- ⚪ May 14
- ⚪ May 21

3. Which shape can you fold to show two equal parts?

- ⚪ **F**
- ⚪ **N**
- ⚪ **U**
- ⚪ **G**

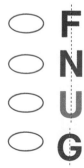

Statistics, Data Analysis, and Probability

4. Bill picks a letter out of the bag. His eyes are closed. Which letter is Bill most likely to pick?

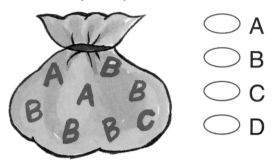

- ⚪ A
- ⚪ B
- ⚪ C
- ⚪ D

5. Lulu picks a square from the bag 4 times. Which shows the tally marks she made to keep track?

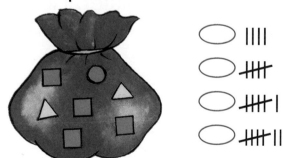

- ⚪ ||||
- ⚪ 卌
- ⚪ 卌|
- ⚪ 卌||

6. Which pet did 6 children vote for?

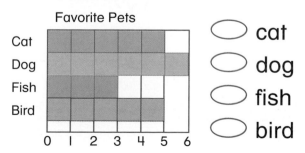

Favorite Pets

- ⚪ cat
- ⚪ dog
- ⚪ fish
- ⚪ bird

Mathematical Reasoning

7. Draw a picture you could use to solve this problem. Tom has 6 toy cars. He buys 2 more. How many does he have now?

8. Maria says the answer to this problem is 4. Is she correct? Tell how you know. Doug has 12 marbles. He gave 8 to his sister. How many does he have left?

_____

_____

9. Betsy has 6 beads on her necklace. She buys 4 more. How many beads does Betsy have now?

_____

Number Sense

10. I am a number between thirty and forty. When you count by 5, you say my name. What number am I?

- ◯ 25
- ◯ 30
- ◯ 35
- ◯ 45

11. Which number comes between 38 and 40?

- ◯ 37
- ◯ 38
- ◯ 39
- ◯ 40

12. Start at the left. Which rabbit is third?

- ◯ red rabbit
- ◯ yellow rabbit
- ◯ blue rabbit
- ◯ green rabbit

Use the Data
Tell an addition story about the pictures.

What You Will Learn
In this chapter you will learn how to:

- Add 2-digit numbers.
- Add money amounts.
- Add three 2-digit numbers.
- Draw pictures to solve problems.

Dear Family,

In Chapter 5, I will learn how to add 2-digit numbers. Here are new vocabulary words and an activity that we can do together.

Add It!

- Make a set of cards with 2-digit numbers on them, such as 25, 34, and 61.

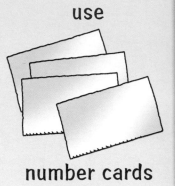

25   34   61

- Ask your child to pick 2 cards.
- Have your child add the two numbers.
- Repeat the activity several times.

use

number cards

paper and pencil

Math Words

addend

```
 14 ← addend
+12 ← addend
 26
```

sum

```
 14
+12
 26 ← sum
```

regroup

12 ones = 1 ten 2 ones

estimate

28 + 42

30 + 40

about 70 ← estimate

Additional activities at
www.mhschool.com/math

Learn

Use a hundred chart. Count on to add 38 + 30.

Start at 38.
Count on 3 tens.
48, 58, 68

1	2	3	4	5	6	7	8	9	10
11	12	13	14	15	16	17	18	19	20
21	22	23	24	25	26	27	28	29	30
31	32	33	34	35	36	37	38	39	40
41	42	43	44	45	46	47	48	49	50
51	52	53	54	55	56	57	58	59	60
61	62	63	64	65	66	67	68	69	70
71	72	73	74	75	76	77	78	79	80
81	82	83	84	85	86	87	88	89	90
91	92	93	94	95	96	97	98	99	100

$38 + 30 = \underline{68}$

Try it   Add. Use the hundred chart.

1. $35 + 20 = \underline{55}$ $27 + 40 = \underline{}$ $50 + 10 = \underline{}$

2. $73 + 20 = \underline{}$ $64 + 30 = \underline{}$ $81 + 10 = \underline{}$

3.

$\begin{array}{r} 30 \\ + 49 \\ \hline \end{array}$ $\begin{array}{r} 28 \\ + 20 \\ \hline \end{array}$ $\begin{array}{r} 70 \\ + 20 \\ \hline \end{array}$ $\begin{array}{r} 40 \\ + 22 \\ \hline \end{array}$ $\begin{array}{r} 29 \\ + 30 \\ \hline \end{array}$ $\begin{array}{r} 10 \\ + 20 \\ \hline \end{array}$

Sum it Up!   How many tens do you count on to add 23 + 20?

Math at Home: Your child added multiples of 10 to 2-digit numbers, for example, 26 + 30.
Activity: Have your child explain how to find the sum for 26 + 30.

one hundred fifty-seven **157**

Practice

Add.

Remember to count on by tens.

4. $54 + 30 = 84$ $60 + 25 = \underline{}$ $28 + 40 = \underline{}$

5. $33 + 40 = \underline{}$ $71 + 20 = \underline{}$ $40 + 19 = \underline{}$

6.
$$\begin{array}{r} 55 \\ +\ 40 \\ \hline \end{array} \qquad \begin{array}{r} 34 \\ +\ 10 \\ \hline \end{array} \qquad \begin{array}{r} 50 \\ +\ 10 \\ \hline \end{array} \qquad \begin{array}{r} 23 \\ +\ 60 \\ \hline \end{array} \qquad \begin{array}{r} 44 \\ +\ 30 \\ \hline \end{array} \qquad \begin{array}{r} 56 \\ +\ 10 \\ \hline \end{array}$$

7.
$$\begin{array}{r} 37 \\ +\ 50 \\ \hline \end{array} \qquad \begin{array}{r} 70 \\ +\ 18 \\ \hline \end{array} \qquad \begin{array}{r} 40 \\ +\ 20 \\ \hline \end{array} \qquad \begin{array}{r} 14 \\ +\ 70 \\ \hline \end{array} \qquad \begin{array}{r} 22 \\ +\ 50 \\ \hline \end{array} \qquad \begin{array}{r} 30 \\ +\ 60 \\ \hline \end{array}$$

8.
$$\begin{array}{r} 82 \\ +\ 10 \\ \hline \end{array} \qquad \begin{array}{r} 60 \\ +\ 20 \\ \hline \end{array} \qquad \begin{array}{r} 43 \\ +\ 50 \\ \hline \end{array} \qquad \begin{array}{r} 27 \\ +\ 30 \\ \hline \end{array} \qquad \begin{array}{r} 20 \\ +\ 68 \\ \hline \end{array} \qquad \begin{array}{r} 70 \\ +\ 10 \\ \hline \end{array}$$

Algebra & functions

Find each missing addend.

9. $24 + \boxed{} = 54$

$\boxed{} + 10 = 65$

$3 + \boxed{} = 43$

$\boxed{} + 20 = 82$

10. $62 + \boxed{} = 92$

$\boxed{} + 43 = 63$

$5 + \boxed{} = 75$

$\boxed{} + 20 = 67$

Add.

1. 4 + 4 = ____

2. 8 + 3 = ____

3. 7 + 6 = ____

4. 1 + 1 = ____

5. 3 + 4 = ____

6. 2 + 7 = ____

7. 8 + 5 = ____

8. 5 + 1 = ____

9. 3 + 8 = ____

10. 8 + 4 = ____

11. 3 + 7 = ____

12. 8 + 0 = ____

13. 9 + 8 = ____

14. 7 + 5 = ____

15. 2 + 2 = ____

16. 6 + 9 = ____

17. 9 + 5 = ____

18. 5 + 3 = ____

19. 4 + 5 = ____

20. 9 + 2 = ____

21. 6 + 3 = ____

22. 1 + 2 = ____

23. 6 + 7 = ____

24. 0 + 5 = ____

McGraw-Hill School Division

Math at Home: Your child practiced addition facts.
Activity: Cover the answers with a paper strip. Time your child as he or she writes the
answers. You can repeat daily to help your child recall facts.

one hundred fifty-nine **159**

Facts Practice: Addition

Add.

1. $\begin{array}{r} 8 \\ +\ 6 \\ \hline \end{array}$ $\begin{array}{r} 6 \\ +\ 4 \\ \hline \end{array}$ $\begin{array}{r} 2 \\ +\ 3 \\ \hline \end{array}$ $\begin{array}{r} 7 \\ +\ 9 \\ \hline \end{array}$ $\begin{array}{r} 9 \\ +\ 4 \\ \hline \end{array}$ $\begin{array}{r} 8 \\ +\ 7 \\ \hline \end{array}$ $\begin{array}{r} 2 \\ +\ 7 \\ \hline \end{array}$

2. $\begin{array}{r} 9 \\ +\ 1 \\ \hline \end{array}$ $\begin{array}{r} 7 \\ +\ 6 \\ \hline \end{array}$ $\begin{array}{r} 3 \\ +\ 3 \\ \hline \end{array}$ $\begin{array}{r} 7 \\ +\ 0 \\ \hline \end{array}$ $\begin{array}{r} 9 \\ +\ 6 \\ \hline \end{array}$ $\begin{array}{r} 6 \\ +\ 8 \\ \hline \end{array}$ $\begin{array}{r} 8 \\ +\ 0 \\ \hline \end{array}$

3. $\begin{array}{r} 5 \\ +\ 6 \\ \hline \end{array}$ $\begin{array}{r} 8 \\ +\ 8 \\ \hline \end{array}$ $\begin{array}{r} 7 \\ +\ 3 \\ \hline \end{array}$ $\begin{array}{r} 6 \\ +\ 2 \\ \hline \end{array}$ $\begin{array}{r} 7 \\ +\ 8 \\ \hline \end{array}$ $\begin{array}{r} 3 \\ +\ 4 \\ \hline \end{array}$ $\begin{array}{r} 2 \\ +\ 5 \\ \hline \end{array}$

4. $\begin{array}{r} 8 \\ +\ 2 \\ \hline \end{array}$ $\begin{array}{r} 6 \\ +\ 6 \\ \hline \end{array}$ $\begin{array}{r} 8 \\ +\ 9 \\ \hline \end{array}$ $\begin{array}{r} 5 \\ +\ 9 \\ \hline \end{array}$ $\begin{array}{r} 6 \\ +\ 1 \\ \hline \end{array}$ $\begin{array}{r} 9 \\ +\ 9 \\ \hline \end{array}$ $\begin{array}{r} 7 \\ +\ 3 \\ \hline \end{array}$

5. $\begin{array}{r} 0 \\ +\ 1 \\ \hline \end{array}$ $\begin{array}{r} 9 \\ +\ 8 \\ \hline \end{array}$ $\begin{array}{r} 8 \\ +\ 0 \\ \hline \end{array}$ $\begin{array}{r} 1 \\ +\ 0 \\ \hline \end{array}$ $\begin{array}{r} 5 \\ +\ 5 \\ \hline \end{array}$ $\begin{array}{r} 1 \\ +\ 9 \\ \hline \end{array}$ $\begin{array}{r} 8 \\ +\ 7 \\ \hline \end{array}$

6. $\begin{array}{r} 7 \\ +\ 4 \\ \hline \end{array}$ $\begin{array}{r} 2 \\ +\ 1 \\ \hline \end{array}$ $\begin{array}{r} 6 \\ +\ 7 \\ \hline \end{array}$ $\begin{array}{r} 2 \\ +\ 2 \\ \hline \end{array}$ $\begin{array}{r} 0 \\ +\ 0 \\ \hline \end{array}$ $\begin{array}{r} 1 \\ +\ 1 \\ \hline \end{array}$ $\begin{array}{r} 8 \\ +\ 2 \\ \hline \end{array}$

7. $\begin{array}{r} 6 \\ +\ 0 \\ \hline \end{array}$ $\begin{array}{r} 7 \\ +\ 7 \\ \hline \end{array}$ $\begin{array}{r} 8 \\ +\ 1 \\ \hline \end{array}$ $\begin{array}{r} 3 \\ +\ 9 \\ \hline \end{array}$ $\begin{array}{r} 1 \\ +\ 7 \\ \hline \end{array}$ $\begin{array}{r} 9 \\ +\ 0 \\ \hline \end{array}$ $\begin{array}{r} 4 \\ +\ 1 \\ \hline \end{array}$

Name _____

Learn

You can use addition facts.

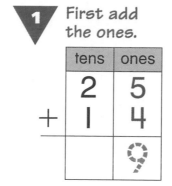

25 children visited the library on Tuesday. 14 children visited the library on Thursday. How many children visited the library in all?

Add to find how many in all.

1 First add the ones.

tens	ones
2	5
+ 1	4
	9

2 Then add the tens.

tens	ones
2	5
+ 1	4
3	9

_39_ children in all

Try it   Add.

1.

tens	ones
4	7
+ 2	1
6	8

tens	ones
1	5
+ 3	2

tens	ones
3	6
+	2

tens	ones
2	5
+ 3	1

2.

```
  12      34      71      22      81      57
+ 43    +  2    + 18    +  5    + 14    + 10
```

3.

```
  31      45      62      42      36      44
+  7    + 13    +  5    + 30    +  3    + 21
```

What addition facts can help you add 32 + 26?

🏠 **Math at Home:** Your child used basic facts to add tens and ones.
Activity: Have your child add 43 + 24.

Add.

Add the ones first.

4.

tens	ones
2	5
+ 4	2
6	7

tens	ones
6	4
+ 2	3

tens	ones
4	2
+	7

tens	ones
2	5
+ 3	1

5.

12	23	11	32	41	62
+ 12	+ 5	+ 46	+ 24	+ 3	+ 17

6.

32	73	43	53	72	54
+ 5	+ 12	+ 24	+ 4	+ 26	+ 21

7.

12	24	44	46	34	54
+ 16	+ 2	+ 32	+ 13	+ 11	+ 32

8.

41	18	53	50	77	31
+ 7	+ 31	+ 35	+ 18	+ 2	+ 57

9.

62	43	26	55	63	22
+ 10	+ 6	+ 41	+ 1	+ 5	+ 3

10. What could the next number be?
Explain the pattern.

24	34	44	54	____
47	57	67	77	____
51	41	31	21	____
98	88	78	68	____

Name _____

Learn

You may need to regroup.

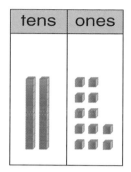

Use tens and ones models to add 27 + 5.

1 Show 27.

tens	ones

2 tens 7 ones

2 Add 5 ones.

tens	ones

2 tens 12 ones

3 Regroup 10 ones as 1 ten.

tens	ones

3 tens 2 ones in all

27 + 5 = __32__

Try it

Use ▭▭▭ and ▪ .

	Add.	Do you need to regroup?	How many in all?
1.	36 + 5	(yes) no	41
2.	28 + 3	yes no	
3.	25 + 4	yes no	
4.	43 + 9	yes no	

Workspace

Sum it Up

How do you know when to regroup ones?

Math at Home: Your child added two numbers, deciding each time if regrouping was needed.
Activity: Have your child use straws to represent tens and beans to represent ones to show you how to add 18 + 6.

Practice

Use ▭▭▭▭ and ▪ .

Regroup when you need to.

	Add.	Do you need to regroup?	How many in all?
5.	27 + 6	(yes) no	33
6.	35 + 0	yes no	
7.	56 + 8	yes no	
8.	48 + 9	yes no	
9.	23 + 6	yes no	
10.	71 + 4	yes no	
11.	34 + 8	yes no	
12.	12 + 9	yes no	
13.	23 + 5	yes no	
14.	39 + 1	yes no	
15.	42 + 8	yes no	

Workspace

Critical Thinking — Journal

16. Draw a picture. Show the sum of 35 + 10 using tens and ones.

Learn

Add 25 + 36.

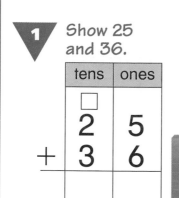

1 Show 25 and 36.

tens	ones
☐ 2	5
+ 3	6

2 Add the ones. Regroup if you need to.

tens	ones
☐ 2	5
+ 3	6
	1

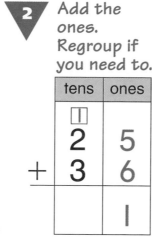

3 Add the tens.

tens	ones
☐ 2	5
+ 3	6
6	1

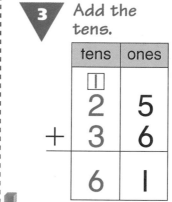

25 + 36 = _61_

Try it

Add. Regroup if you need to.

1.

tens	ones
☐ 4	3
+ 2	8
7	1

tens	ones
☐ 1	9
+ 3	6

tens	ones
☐ 2	2
+ 1	7

tens	ones
☐ 3	4
+ 4	9

2.

tens	ones
☐ 1	5
+ 5	9

tens	ones
☐ 2	3
+	2

tens	ones
☐ 1	4
+	7

tens	ones
☐ 2	1
+ 4	3

How do you show regrouping 10 ones as 1 ten in 26 + 14?

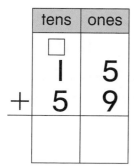

Math at Home: Your child added two numbers, regrouping 10 ones as 1 ten when needed.
Activity: Ask your child to show you how to add 37 and 46.

Decide if you need to regroup.

3.

tens	ones
□ 4	1
+ 1	7
5	8

tens	ones
□ 2	5
+	9

tens	ones
□ 6	1
+ 2	4

tens	ones
□ 3	3
+ 5	9

4.

tens	ones
□ 2	8
+	5

tens	ones
□ 3	6
+ 1	0

tens	ones
□ 2	1
+	8

tens	ones
□ 4	2
+ 1	6

5.

tens	ones
□ 1	8
+ 2	2

tens	ones
□ 3	9
+	2

tens	ones
□ 2	6
+ 4	8

tens	ones
□ 5	3
+	7

Spiral Review and Test Prep

Choose the correct answer.

6. 37 + 48 = ☐

- ⬭ 65
- ⬭ 75
- ⬭ 85
- ⬭ 95

7. I am a number between fifty and sixty. When you count by 5, you say my name. What number am I?

- ⬭ 35
- ⬭ 45
- ⬭ 54
- ⬭ 55

Learn

Use tens and ones models to add.

Do you need to regroup?

There are 12 ones. Regroup 10 ones as 1 ten.

tens	ones
⬚ 4	8
+	4
5	2

There are 9 ones. Do not regroup.

tens	ones
⬚ 2	8
+ 3	1
5	9

Try it

Add. You can use tens and ones models.

1.

tens	ones
⬚ 2	6
+	8

tens	ones
⬚ 1	5
+ 3	4

tens	ones
⬚ 1	8
+ 2	6

tens	ones
4	4
+	5

2.

tens	ones
⬚ 1	6
+ 6	2

tens	ones
⬚ 1	8
+ 3	4

tens	ones
⬚ 8	2
+	6

tens	ones
⬚ 4	9
+	5

Sum it Up! How can you tell when you need to regroup?

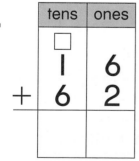

Math at Home: Your child decided whether to regroup, then added.
Activity: Ask your child to show you addition that requires regrouping and addition that does not require regrouping.

Practice

Add. You can use tens and ones models.

Decide if you need to regroup.

3.

tens	ones
□ 3	5
+	8

tens	ones
□ 2	2
+ 1	4

tens	ones
□ 1	7
+ 2	3

tens	ones
□ 1	0
+	6

4.

tens	ones
□ 2	8
+ 3	5

tens	ones
□ 5	3
+	6

tens	ones
□ 2	8
+ 4	0

tens	ones
□ 3	9
+	5

5.

tens	ones
□ 2	6
+	3

tens	ones
□ 2	5
+ 4	8

tens	ones
□ 1	8
+	9

tens	ones
□ 1	4
+ 2	5

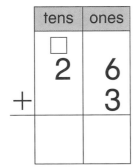 **Problem Solving**

Data.

6. Use the chart.
How many tickets were sold
on Monday and Friday? _____ tickets

Were more tickets sold on
Wednesday or Friday? _____

Museum Tickets Sold	
Monday	17
Wednesday	23
Friday	15

Name _____

Learn

18 children visit the post office.
5 more children join the group.
How many children visit the post office?

 1 Add the ones.
Regroup if you need to.

1

```
  1 8
+   5
─────
    3
```

 2 Add the tens.

1

```
  1 8
+   5
─────
  2 3
```

__23__ children visit the post office.

Do you need to regroup?

Try it

Add. You can use tens and ones models.

1.
```
  2 7      4 3      7 3      3 9      5 5
+   6    +   5    + 2 4    + 3 8    + 1 3
─────
  3 3
```

2.
```
  3 6      7 8      6 5      1 7      2 3
+ 1 0    +   5    + 2 6    + 3 9    +   9
```

Sum it Up

How do you add two 2-digit numbers?

 **Math at Home:** Your child practiced 2-digit addition with and without regrouping.
Activity: Ask your child to show you how to add 38 + 47.

Practice

Add. You can use tens and ones models.

Regroup if you need to.

3.

□1	□	□	□	□
16	24	42	54	38
+ 5	+ 4	+ 31	+ 26	+ 3

21

4.

□	□	□	□	□
18	26	37	25	69
+ 3	+ 7	+ 43	+ 4	+ 10

5.

□	□	□	□	□
37	29	41	44	28
+ 11	+ 13	+ 0	+ 9	+ 16

6.

□	□	□	□	□
19	21	62	26	85
+ 27	+ 8	+ 9	+ 21	+ 8

7. Rewrite. Then add.

45 + 21 + 45
          21
          ——
          66

26 + 17 + □
          □
          ——
          □

44 + 39 + □
          □
          ——
          □

Name _____

Important and Unimportant Information

Reading Skill   You can identify important information to help you solve problems.

It's the Soccer Playoff! There are 18 children watching the playoffs. There are 35 parents watching. The score is 3 to 1. The Bluebirds are winning.

How many people are watching the game?

1. Does the information help you to solve the problem?
   Circle yes or no.

   - The Bluebirds are winning.   yes   (no)
   - The score is 3 to 1.   yes   no
   - There are 18 children watching.   yes   no
   - There are 35 parents watching.   yes   no

2. Solve the problem. Write a number sentence.

_____

 Math at Home: Your child identified important information to help him or her solve problems.
Activity: Have your child tell you what information is important in this story.

one hundred seventy-one

Solve.

There are 12 blue houses on Maple Street.
There are 12 green houses.

There are 9 children in the neighborhood. 6 children are in the second grade.

How many houses are on Maple Street?

6. Does the information help you to solve the problem? Circle yes or no.

- There are 12 green houses.   yes   no

- There are 9 children in the neighborhood.   yes   no

- There are 12 blue houses.   yes   no

- 6 children are in second grade.   yes   no

7. Solve the problem. Write a number sentence. Then label the parts of the number sentence.

_____

Critical Thinking   *Journal*

8. Use the story. Write a problem.

_____

Name _____

Add.

1.

$$\begin{array}{r} 40 \\ + 22 \\ \hline \end{array}$$
$$\begin{array}{r} 60 \\ + 34 \\ \hline \end{array}$$
$$\begin{array}{r} 45 \\ + 30 \\ \hline \end{array}$$
$$\begin{array}{r} 50 \\ + 22 \\ \hline \end{array}$$
$$\begin{array}{r} 56 \\ + 20 \\ \hline \end{array}$$
$$\begin{array}{r} 18 \\ + 40 \\ \hline \end{array}$$

2.

$$\begin{array}{r} 31 \\ + 6 \\ \hline \end{array}$$
$$\begin{array}{r} 72 \\ + 26 \\ \hline \end{array}$$
$$\begin{array}{r} 47 \\ + 11 \\ \hline \end{array}$$
$$\begin{array}{r} 53 \\ + 6 \\ \hline \end{array}$$
$$\begin{array}{r} 16 \\ + 22 \\ \hline \end{array}$$
$$\begin{array}{r} 24 \\ + 32 \\ \hline \end{array}$$

Add. Regroup if you need to.

3.

☐
$$\begin{array}{r} 29 \\ + 4 \\ \hline \end{array}$$

☐
$$\begin{array}{r} 62 \\ + 6 \\ \hline \end{array}$$

☐
$$\begin{array}{r} 32 \\ + 16 \\ \hline \end{array}$$

☐
$$\begin{array}{r} 27 \\ + 46 \\ \hline \end{array}$$

4.

☐
$$\begin{array}{r} 23 \\ + 8 \\ \hline \end{array}$$

☐
$$\begin{array}{r} 31 \\ + 8 \\ \hline \end{array}$$

☐
$$\begin{array}{r} 72 \\ + 16 \\ \hline \end{array}$$

☐
$$\begin{array}{r} 49 \\ + 11 \\ \hline \end{array}$$

Compare the numbers. Write >, <, or =.

1. 35 ◯ 53   19 ◯ 19   43 ◯ 38

2. 84 ◯ 46   29 ◯ 33   24 ◯ 24

3. 68 ◯ 68   61 ◯ 16   12 ◯ 21

Write each number.

4. nineteen _____   90 _____

5. ninety-eight _____   74 _____

6. seventy-five _____   fifty-nine _____

Circle the coins that show the amount.

7. 45¢

TECHNOLOGY LINK

Use Place Value Models to Add
- Choose place value.
- Choose a mat to add.
- Stamp out 37 and 45.
- Click on +.
- Trade up as needed.
- What is the sum?

   1. Choose place value. Stamp out 25 + 39.
   What is the sum?

   2. Stamp out other numbers. Tell the sum.

For more practice use Math Traveler™.

Name _____

Draw a Picture

What do I know?

What do I need to know?

Read ▶ 15 second graders visit the nature museum on Monday.
22 second graders visit on Tuesday. How many second graders visit the museum in all?

Plan ▶ I can draw tally marks to help me.

Solve ▶ Monday   ~~IIII~~ ~~IIII~~ IIII

Tuesday   ~~IIII~~ ~~IIII~~ ~~IIII~~ ~~IIII~~ II   **37 second graders**

Look Back ▶ Does my answer make sense? Why?

Try it   Draw a picture. Solve.

Workspace

1. 11 second graders skated to the park on Saturday. 17 second graders walked to the park on Saturday. How many second graders came to the park in all on Saturday?

   _____ second graders

**Sum it Up**   How can drawing a picture help you solve a problem?

Math at Home: Your child drew pictures to solve problems.
Activity: Tell your child a simple problem and have him or her draw a picture to solve the problem.

Draw a picture. Solve.

Workspace

2. 10 children went to the aquarium on Thursday. 5 children went on Friday. How many children went on the two days?

_____ children

3. 16 children take the first bus to the movies. 18 children take the second bus. How many children are on the two buses?

_____ children

4. Carry is at the library. She finds 8 books on sharks and 12 books on horses. How many books did she find in all?

_____ books

5. Look at the picture.
Use numbers.
Write a word problem.

_____

_____

Algebra & functions **Check Addition**

Learn

The children at Madison School gave 16 animal books to the library. They also gave 28 coloring books. How many books did the children give to the library?

You can check addition by adding the numbers in a different order.

Add.

$$\begin{array}{r} 16 \\ + 28 \\ \hline 44 \end{array}$$

Add the numbers in a different order to check.

$$\begin{array}{r} 28 \\ + 16 \\ \hline 44 \end{array}$$

44 books

Try it   Add. Check by adding in a different order.

1.
$$\begin{array}{r} 17 \\ + 21 \\ \hline 38 \end{array}$$
$$\begin{array}{r} 21 \\ + 17 \\ \hline 38 \end{array}$$

$$\begin{array}{r} 12 \\ + 29 \\ \hline \end{array}$$
$$\begin{array}{r} \\ + \\ \hline \end{array}$$

$$\begin{array}{r} 47 \\ + 13 \\ \hline \end{array}$$
$$\begin{array}{r} \\ + \\ \hline \end{array}$$

2.
$$\begin{array}{r} 48 \\ + 37 \\ \hline \end{array}$$
$$\begin{array}{r} \\ + \\ \hline \end{array}$$

$$\begin{array}{r} 9 \\ + 19 \\ \hline \end{array}$$
$$\begin{array}{r} \\ + \\ \hline \end{array}$$

$$\begin{array}{r} 65 \\ + 8 \\ \hline \end{array}$$
$$\begin{array}{r} \\ + \\ \hline \end{array}$$

Sum it Up   How do you know that you have the correct answer when you check addition?

Math at Home: Your child learned to check addition by reversing the order.
Activity: Ask your child to add 32 + 46. Then have him or her show you how to check the answer.

one hundred seventy-seven

Check by adding in a different order.

3.

$$\begin{array}{r} 34 \\ + 19 \\ \hline 53 \end{array} \qquad \begin{array}{r} 19 \\ + 34 \\ \hline 53 \end{array} \qquad \begin{array}{r} 7 \\ + 37 \\ \hline \end{array} \qquad + \underline{} \qquad \begin{array}{r} 47 \\ + 19 \\ \hline \end{array} \qquad + \underline{}$$

4.

$$\begin{array}{r} 67 \\ + 28 \\ \hline \end{array} \qquad + \underline{} \qquad \begin{array}{r} 25 \\ + 30 \\ \hline \end{array} \qquad + \underline{} \qquad \begin{array}{r} 48 \\ + 14 \\ \hline \end{array} \qquad + \underline{}$$

5.

$$\begin{array}{r} 18 \\ + 17 \\ \hline \end{array} \qquad + \underline{} \qquad \begin{array}{r} 63 \\ + 34 \\ \hline \end{array} \qquad + \underline{} \qquad \begin{array}{r} 18 \\ + 42 \\ \hline \end{array} \qquad + \underline{}$$

6.

$$\begin{array}{r} 22 \\ + 38 \\ \hline \end{array} \qquad + \underline{} \qquad \begin{array}{r} 6 \\ + 19 \\ \hline \end{array} \qquad + \underline{} \qquad \begin{array}{r} 33 \\ + 46 \\ \hline \end{array} \qquad + \underline{}$$

Spiral Review and Test Prep

Choose the best answer.

7. Which names the same number as $47 + 23$?

- ○ $47 + 47$
- ○ $74 + 32$
- ○ $23 + 47$
- ○ $23 + 23$

8. Which of these is not a doubles fact?

- ○ $6 + 6$
- ○ $5 + 4$
- ○ $5 + 5$
- ○ $7 + 7$

Learn

Kim is going to swim class. She bikes 12 minutes on Elm Street. She bikes 19 minutes on Pike Street. How many minutes does Kim bike in all?

Rewrite the numbers to the nearest 10.

$12 + 19 = 31$ Kim bikes for 31 minutes.

You can estimate to see if the answer is reasonable.

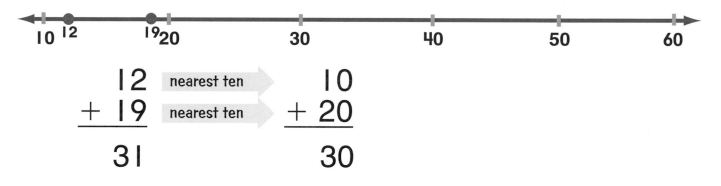

$$
\begin{array}{r}
12 \\
+ 19 \\
\hline
31
\end{array}
\quad
\begin{array}{c}
\text{nearest ten} \rightarrow \\
\text{nearest ten} \rightarrow
\end{array}
\quad
\begin{array}{r}
10 \\
+ 20 \\
\hline
30
\end{array}
$$

31 is close to 30.
The answer is reasonable.

Try it

Add. Estimate to see if your answer is reasonable.

1.
$$
\begin{array}{r}
41 \\
+ 58 \\
\hline
99
\end{array}
\qquad
\begin{array}{r}
40 \\
+ 60 \\
\hline
100
\end{array}
\qquad
\begin{array}{r}
37 \\
+ 49 \\
\hline
\end{array}
\quad
\begin{array}{r}
\\
+ \underline{}
\end{array}
\qquad
\begin{array}{r}
29 \\
+ 62 \\
\hline
\end{array}
\quad
\begin{array}{r}
\\
+ \underline{}
\end{array}
$$

Sum it Up How can you use estimation to see if your answer is reasonable?

 **Math at Home:** Your child estimated sums by finding the nearest ten.
Activity: Ask your child to explain how to estimate 38 + 27.

Add. Estimate to see if your answer is reasonable.

2.
$$\begin{array}{r} 22 \\ + 49 \\ \hline \end{array}$$
$$\begin{array}{r} 20 \\ + 50 \\ \hline 70 \end{array}$$
71

$$\begin{array}{r} 17 \\ + 62 \\ \hline \end{array}$$
$+\underline{}$

$$\begin{array}{r} 38 \\ + 41 \\ \hline \end{array}$$
$+\underline{}$

3.
$$\begin{array}{r} 48 \\ + 39 \\ \hline \end{array}$$
$+\underline{}$

$$\begin{array}{r} 21 \\ + 59 \\ \hline \end{array}$$
$+\underline{}$

$$\begin{array}{r} 39 \\ + 51 \\ \hline \end{array}$$
$+\underline{}$

4.
$$\begin{array}{r} 52 \\ + 21 \\ \hline \end{array}$$
$+\underline{}$

$$\begin{array}{r} 11 \\ + 48 \\ \hline \end{array}$$
$+\underline{}$

$$\begin{array}{r} 69 \\ + 21 \\ \hline \end{array}$$
$+\underline{}$

5.
$$\begin{array}{r} 32 \\ + 48 \\ \hline \end{array}$$
$+\underline{}$

$$\begin{array}{r} 13 \\ + 47 \\ \hline \end{array}$$
$+\underline{}$

$$\begin{array}{r} 29 \\ + 57 \\ \hline \end{array}$$
$+\underline{}$

Problem Solving

6. Jenny is making a bead necklace.
   Draw and color the next two beads
   that would most likely continue her pattern.

Learn

Sam buys marbles for 49¢.
He buys a toy car for 25¢.
How much does he spend
in all?

$$\begin{array}{r} 49¢ \\ + 25¢ \\ \hline 74¢ \end{array}$$

Sam spends 74¢.

Try it   Add.

1.
$$\begin{array}{r} 12¢ \\ + 29¢ \\ \hline 41¢ \end{array}$$
$$\begin{array}{r} 36¢ \\ + 8¢ \\ \hline \end{array}$$
$$\begin{array}{r} 25¢ \\ + 14¢ \\ \hline \end{array}$$
$$\begin{array}{r} 73¢ \\ + 17¢ \\ \hline \end{array}$$
$$\begin{array}{r} 41¢ \\ + 16¢ \\ \hline \end{array}$$
$$\begin{array}{r} 57¢ \\ + 18¢ \\ \hline \end{array}$$

2.
$$\begin{array}{r} 15¢ \\ + 13¢ \\ \hline \end{array}$$
$$\begin{array}{r} 49¢ \\ + 34¢ \\ \hline \end{array}$$
$$\begin{array}{r} 28¢ \\ + 5¢ \\ \hline \end{array}$$
$$\begin{array}{r} 12¢ \\ + 32¢ \\ \hline \end{array}$$
$$\begin{array}{r} 14¢ \\ + 6¢ \\ \hline \end{array}$$
$$\begin{array}{r} 57¢ \\ + 35¢ \\ \hline \end{array}$$

Sum it Up

How is adding money amounts like adding
other numbers?

Math at Home: Your child added money amounts up to 99¢.
Activity: Have your child show you how to add 39¢ + 55¢.

Remember to write ¢.

3.

22¢	35¢	15¢	22¢	72¢	42¢
+ 37¢	+ 8¢	+ 14¢	+ 34¢	+ 19¢	+ 36¢

59¢

4.

11¢	26¢	35¢	34¢	62¢	44¢
+ 24¢	+ 43¢	+ 29¢	+ 8¢	+ 9¢	+ 18¢

5.

58¢	17¢	28¢	56¢	14¢	68¢
+ 32¢	+ 11¢	+ 62¢	+ 9¢	+ 10¢	+ 23¢

6.

25¢	27¢	83¢	34¢	36¢	57¢
+ 39¢	+ 53¢	+ 16¢	+ 7¢	+ 24¢	+ 28¢

Problem Solving

7. Josie buys 2 toys for 38¢ each. How much does she spend?

Workspace

Name _____

Learn

Regroup if you need to.

The children's center put on 3 plays.
34 children were in *Drum Song*.
16 children were in *The Little Pony*.
24 children were in *Chipper's Home Run*.
How many children were in the plays?

1 Add the ones.

```
  1
  34
  16
+ 24
────
   4
```

2 Add the tens.

```
  1
  34
  16
+ 24
────
  74
```

74 children

Try it

Add. You can use tens and ones models.

1.
```
  32       23       14       46       25       23
  18       18        9       26       35       15
+ 25      + 3      + 31     + 17     + 38     + 37
```

2.
```
  15       42       62       23       31       54
  15       32        8       17       19       26
+ 25      +14      + 21     + 32     + 20     +  5
```

Sum it Up

How is adding 3 numbers like adding 2 numbers?

Math at Home: Your child added three 2-digit numbers.
Activity: Have your child explain how to add 24 + 15 + 16.

Practice

Add. You can use tens and ones models.

> You can add tens or doubles first.

3.

44	17	28	37	15	29
12	7	32	52	35	17
+ 26	+ 35	+ 23	+ 3	+ 22	+ 41

4.

32	32	12	36	53	62
22	41	25	21	22	4
+ 37	+ 12	+ 51	+ 2	+ 13	+ 13

5.

18	34	6	43	17	29
21	14	25	10	37	23
+ 4	+ 27	+ 36	+ 23	+ 40	+ 12

6.

18	32	15	71	46	28
5	10	8	11	34	16
+48	+ 21	+ 33	+ 9	+ 6	+ 20

Algebra & functions

Find each missing addend.

7. $32 + 13 + \underline{\hspace{1cm}} = 87$

8. $21 + 36 + \underline{\hspace{1cm}} = 71$

$26 + \underline{\hspace{1cm}} + 16 = 73$

$12 + \underline{\hspace{1cm}} + 65 = 94$

Name _____

Plan a Bus Route

You want to plan a school bus route
from your house to school.
Choose a route.

Use yarn to show the route.
Then draw the route on the map.

1. How did you decide which bus route to take?

_____

_____

_____

2. How many blocks is the bus route you chose?

_____

> **Workspace**

3. **What if** you planned another bus route? How many blocks is this route? Is it shorter or longer than the other route?

_____

_____

_____

_____

> **Workspace**

Name_____

What Do You See in a Tree?

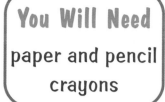

You Will Need

paper and pencil
crayons

What to do

1. Take a walk around your school.

2. Look at different trees.

3. How many different kinds of trees do you see?

4. Look at the tree parts. Notice how the leaves are arranged.

5. How are the leaves alike?

6. How are the leaves different?

Problem Solving • Math and Science

What did you find out?

Draw a picture of your tree. Label the parts.

1. Tell about your tree.

_____

2. How are trees alike?

_____

3. How are trees different?

_____

4. How do you think trees get water? Explain.

_____

_____

5. Collect leaves from two different kinds of trees. Add to find how many leaves you have in all.

_____ leaves.

 Want to do more?

6. Find out about trees in your community. How many different kinds of trees grow where you live?

_____

Workspace

Did You KNOW?

A tree is a plant. A tree has leaves, branches, and roots.

Name _____

Add. Check by adding in a different order.

1.
```
   46          33          27
 + 19    +    + 59    +    + 47    +
```

2.
```
   17     23     45     42      9
 + 59   + 58   + 52   + 29    + 32    +
```

Add.

3.
```
   11¢        40¢        61¢        44¢        25¢
 + 43¢      + 25¢      +  9¢      + 14¢      + 25¢
```

4.
```
   19         60         11         20         17
    7         12         34         33         15
 + 53       + 25       + 48       +  8       +27
```

5. Draw a picture. Solve.

   Jill's class read 12 cat books and
   25 rabbit books. How many books
   did they read altogether? _____ books

6. What if Jill's class read
   18 more books? Write a problem to
   show how many books they read in all. _____

Name_____

That Sums It Up!

- You and your partner take turns.
- Drop your counters on the game board.
- Add the numbers.
- The partner with the greater sum gets 1 point.
- Score 1 point for each correct sum.
- The first player who gets 10 points wins.

43	17	6	29	47	32
36	27	45	14	22	16
50	4	35	47	7	44
29	36	16	45	23	35

Name _____

Language and Math

Complete. Use a word from the list.

1. You have to _____
   when you add 56 and 28.

2. You can _____ sums
   to check.

3. When you add numbers the

   answer is the _____ .

Read these words.

Concepts and Skills

Add.

4.
$$\begin{array}{r} 70 \\ + 20 \\ \hline \end{array} \quad \begin{array}{r} 32 \\ + 66 \\ \hline \end{array} \quad \begin{array}{r} 18¢ \\ + 12¢ \\ \hline \end{array} \quad \begin{array}{r} 27 \\ + 54 \\ \hline \end{array} \quad \begin{array}{r} 19 \\ + 20 \\ \hline \end{array} \quad \begin{array}{r} 18 \\ + 44 \\ \hline \end{array}$$

5.
$$\begin{array}{r} 50 \\ + 20 \\ \hline \end{array} \quad \begin{array}{r} 46¢ \\ + 19¢ \\ \hline \end{array} \quad \begin{array}{r} 23 \\ + 16 \\ \hline \end{array} \quad \begin{array}{r} 18 \\ + 25 \\ \hline \end{array} \quad \begin{array}{r} 16 \\ + 8 \\ \hline \end{array} \quad \begin{array}{r} 32 \\ + 46 \\ \hline \end{array}$$

6.
$$\begin{array}{r} 24 \\ + 58 \\ \hline \end{array} \quad \begin{array}{r} 41 \\ + 8 \\ \hline \end{array} \quad \begin{array}{r} 12¢ \\ + 56¢ \\ \hline \end{array} \quad \begin{array}{r} 49 \\ + 18 \\ \hline \end{array} \quad \begin{array}{r} 22 \\ + 48 \\ \hline \end{array} \quad \begin{array}{r} 12¢ \\ + 17¢ \\ \hline \end{array}$$

Add.

7.
4	28	27	5	31
12	30	50	15	17
+ 10	+ 21	+ 3	+ 22	+ 41

8.
40	23	27	15	29
10	11	52	15	47
+ 26	+ 45	+ 3	+ 34	+ 21

Problem Solving

Draw a picture. Solve.

Workspace

9. 17 second graders went to the dinosaur museum on Thursday. 21 second graders went on Friday. How many second graders went to the dinosaur museum?

_____ second graders

10. Jill sees 10 little dinosaurs in one display. She sees 6 bigger dinosaurs in another display. How many dinosaurs does she see in all?

_____ dinosaurs

Name _____

Add.

 blue ➤ Sums greater than 50. green ➤ Sums less than 50.

1.

$$\begin{array}{r} 20 \\ + 21 \\ \hline \end{array}$$
$$\begin{array}{r} 40 \\ + 28 \\ \hline \end{array}$$
$$\begin{array}{r} 20 \\ + 45 \\ \hline \end{array}$$
$$\begin{array}{r} 50 \\ + 26 \\ \hline \end{array}$$

2.

$$\begin{array}{r} 31 \\ + 8 \\ \hline \end{array}$$
$$\begin{array}{r} 42 \\ + 23 \\ \hline \end{array}$$
$$\begin{array}{r} 25 \\ + 14 \\ \hline \end{array}$$
$$\begin{array}{r} 16 \\ + 3 \\ \hline \end{array}$$

3.

$$\begin{array}{r} 28 \\ + 3 \\ \hline \end{array}$$
$$\begin{array}{r} 52 \\ + 7 \\ \hline \end{array}$$
$$\begin{array}{r} 26 \\ + 12 \\ \hline \end{array}$$
$$\begin{array}{r} 39 \\ + 46 \\ \hline \end{array}$$

4.

$$\begin{array}{r} 78¢ \\ + 19¢ \\ \hline \end{array}$$
$$\begin{array}{r} 25¢ \\ + 45¢ \\ \hline \end{array}$$
$$\begin{array}{r} 69¢ \\ + 29¢ \\ \hline \end{array}$$
$$\begin{array}{r} 0¢ \\ + 68¢ \\ \hline \end{array}$$

5.

$$\begin{array}{r} 33 \\ 21 \\ + 24 \\ \hline \end{array}$$
$$\begin{array}{r} 16 \\ 5 \\ + 35 \\ \hline \end{array}$$
$$\begin{array}{r} 18 \\ 52 \\ + 23 \\ \hline \end{array}$$
$$\begin{array}{r} 27 \\ 42 \\ + 3 \\ \hline \end{array}$$

Enrichment

Find the missing digit.

Subtract 8 − 2 to find the missing digit.

```
  5 2
+ 2 6
─────
  7 8
```

1.

```
  3 0        □ 0        2 □        4 □        □ 0
+ 4 □      + 6 □      + □ 0      + 1 0      + 2 □
─────      ─────      ─────      ─────      ─────
  7 5        7 3        4 4        5 2        6 1
```

2.

```
  □ 0        □ 0        3 □        □ 0        2 0
+ 2 2      + 3 □      + □ 5      + 2 □      + □ 6
─────      ─────      ─────      ─────      ─────
  6 2        9 4        7 5        7 7        7 6
```

3.

```
  □ 1        7 2        4 7        5 □        □ □
+ 5 6      + 2 □      + □ 1      + 2 5      + □ 2
─────      ─────      ─────      ─────      ─────
  8 7        9 8        5 □        □ 8        3 8
```

4.

```
  □ 6        1 7        1 8        1 9        4 □
    7          □          7          □          1 8
+   8      +   8      +   8      +   8      +   5
─────      ─────      ─────      ─────      ─────
  3 1        3 2        □ 3        3 4        6 3
```

Chapter Test

Name _____

Add.

1. 30   40   20   24   69   38
     + 40   + 25   + 6   + 34   + 4   + 43

2. 21   23   33   18   61¢   7
     + 27   + 7   + 25   + 44   + 29¢   + 32

3.   33   26   42¢
                            18   48   2¢
    16¢   25   30¢   + 20   + 2   + 16¢
    + 7¢   + 43   + 29¢

Draw a picture. Solve.

| **Workspace** |

4. 10 boys and 12 girls visit City Hall. How many children visit City Hall?

_____ children

5. 4 buses are parked at the station. 3 more buses arrive. How man buses are there in all?

_____ buses

SECOND GRADE CLASSES

TEACHER	NUMBER OF STUDENTS
MRS. REMY	21
MR. GREGORY	18
MRS. NAKANO	17
MR. JONES	22

Use the Table

1. Jeff added to find the total number of students in two classes. He did not have to regroup. Which two classes might Jeff have chosen?

_____

2. Nena added to find the total number of students in two classes. She regrouped. Which two classes might Nena have chosen?

_____

3. Ned added to find the total number of students in two classes. The total was 35. Which two classes did Ned choose?

$+$

_____

4. How many children are in the second grade? _____ children

Name _____

Choose the correct answer.

Mathematical Reasoning

Mark your answer.

1. Kevin has 17 baseball cards. He buys 7 more. How many cards does he have now?

- ⬭ 7
- ⬭ 17
- ⬭ 14
- ⬭ 24

2. Gloria has 95¢. She buys an apple for 50¢. How much change should she receive?

- ⬭ 5¢
- ⬭ 25¢
- ⬭ 40¢
- ⬭ 45¢

Tell how you solved this problem.

_____

_____

3. What number comes between 52 and 54?

- ⬭ 51
- ⬭ 53
- ⬭ 54
- ⬭ 55

Number Sense

4. Which statement is true?

- ⬭ 62 < 57
- ⬭ 62 = 57
- ⬭ 62 > 57
- ⬭ 57 = 62

5. How many  are there in all?

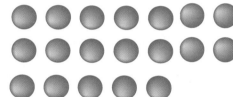

- ⬭ 4
- ⬭ 14
- ⬭ 24
- ⬭ 34

6.
$$\begin{array}{r} 20 \\ + 30 \\ \hline \end{array}$$

- ⬭ 10
- ⬭ 20
- ⬭ 30
- ⬭ 50

McGraw-Hill School Division

Algebra and Functions

7. Which number sentence can you use to solve this problem?

Lois played 3 games. Pam played 5 games. How many games did they play altogether?

○ $3 + 3 = 6$
○ $5 - 3 = 2$
○ $5 - 2 = 3$
○ $3 + 5 = 8$

8. Which sign belongs in the box?

$9 \boxed{} 5 = 4$

○ $=$
○ $+$
○ $-$
○ $\times$

9. 8 frogs are at the pond. 3 hopped away. Which shows how many frogs are left?

○ $3 + 8 = 11$
○ $8 - 5 = 3$
○ $8 - 3 = 5$
○ $11 + 3 = 14$

Measurement and Geometry

10. What could the next shape be in this pattern?

○
○
○
○ ●

11. Use your inch ruler.

About how many inches long is the pencil?

○ I inch
○ 2 inches
○ 3 inches
○ 4 inches

12. What time is shown on the clock?

○ 4:00
○ 4:30
○ 5:30
○ 6:00

Subtract 2-Digit Numbers

Pumpkins

Carrots

theme
Seasons

Use the Data

Tell a subtraction story about the garden.

What You Will Learn

In this chapter you will learn how to:

- Subtract 2-digit numbers.

- Subtract money amounts.

- Add to check subtraction.

- Choose the operation to solve problems.

one hundred ninety-nine **199**

Math at Home

Dear Family,

In Chapter 6, I will learn 2-digit subtraction. Here are vocabulary words and an activity that we can do together.

What's the Difference?

- Make a set of 10 cards with 2-digit numbers, such as 86, 68, 23, and 12.

use

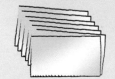

10 index cards

- Ask your child to pick two cards.
- Ask which number is greater.
- Have your child subtract the lesser number from the greater number.
- Repeat with the remaining cards.

Math Words

regroup

1 ten 4 ones = 14 ones

difference

$$
\begin{array}{r}
47 \\
-28 \\
\hline
19
\end{array}
$$
← difference

Additional activities at
www.mhschool.com/math

Learn

Use a hundred chart.
Count back to subtract 48 − 20.

Start at 48.
Count back 2 tens.
38, 28.

1	2	3	4	5	6	7	8	9	10
11	12	13	14	15	16	17	18	19	20
21	22	23	24	25	26	27	28	29	30
31	32	33	34	35	36	37	38	39	40
41	42	43	44	45	46	47	48	49	50
51	52	53	54	55	56	57	58	59	60
61	62	63	64	65	66	67	68	69	70
71	72	73	74	75	76	77	78	79	80
81	82	83	84	85	86	87	88	89	90
91	92	93	94	95	96	97	98	99	100

$48 - 20 = \underline{28}$

Try it

Subtract. Use the hundred chart.

1. $67 - 30 = \underline{37}$ $42 - 20 = \underline{}$ $70 - 10 = \underline{}$

2.
$$\begin{array}{r} 82 \\ -30 \\ \hline \end{array} \qquad \begin{array}{r} 95 \\ -40 \\ \hline \end{array} \qquad \begin{array}{r} 60 \\ -30 \\ \hline \end{array} \qquad \begin{array}{r} 83 \\ -20 \\ \hline \end{array} \qquad \begin{array}{r} 20 \\ -10 \\ \hline \end{array} \qquad \begin{array}{r} 32 \\ -10 \\ \hline \end{array}$$

 How many tens do you count back to subtract 52 − 30?

 Math at Home: Your child subtracted a multiple of 10 from 2-digit numbers; for example 76 - 30.
Activity: Have your child explain how to find the difference for 76 - 30.

two hundred one **201**

Remember to count back by tens.

3. 60 – 20 = ____ 72 – 20 = ____ 93 – 20 = ____

4. 39 – 10 = ____ 81 – 50 = ____ 36 – 10 = ____

5. 54 – 30 = ____ 40 – 30 = ____ 75 – 20 = ____

6.
```
  34        88        77        55        70        63
– 20      – 60      – 50      – 10      – 20      – 40
```

7.
```
  74        83        60        59        80        51
– 40      – 10      – 10      – 30      – 40      – 40
```

8.
```
  90        82        49        30        76        68
– 30      – 40      – 10      – 10      – 20      – 30
```

Algebra & functions Find each missing addend.

9. 84 – ☐ = 54   ☐ – 20 = 54   67 – ☐ = 17

Subtract.

1. $15 - 6 =$ _____

2. $11 - 3 =$ _____

3. $9 - 4 =$ _____

4. $8 - 4 =$ _____

5. $8 - 0 =$ _____

6. $15 - 9 =$ _____

7. $12 - 6 =$ _____

8. $18 - 9 =$ _____

9. $16 - 8 =$ _____

10. $12 - 3 =$ _____

11. $14 - 7 =$ _____

12. $9 - 9 =$ _____

13. $8 - 3 =$ _____

14. $10 - 6 =$ _____

15. $12 - 7 =$ _____

16. $17 - 9 =$ _____

17. $11 - 5 =$ _____

18. $10 - 7 =$ _____

19. $16 - 9 =$ _____

20. $17 - 8 =$ _____

21. $15 - 8 =$ _____

22. $15 - 7 =$ _____

23. $13 - 6 =$ _____

24. $14 - 9 =$ _____

McGraw-Hill School Division

Math at Home: Your child practiced subtraction facts.
Activity: Cover the answers with a paper strip. Time your child as he or she writes the answers again. You can repeat daily to help your child recall the facts quickly.

Facts Practice: Subtraction

Subtract.

1.
$$\begin{array}{r} 15 \\ -\ 8 \\ \hline \end{array}$$
$$\begin{array}{r} 12 \\ -\ 9 \\ \hline \end{array}$$
$$\begin{array}{r} 17 \\ -\ 9 \\ \hline \end{array}$$
$$\begin{array}{r} 10 \\ -\ 4 \\ \hline \end{array}$$
$$\begin{array}{r} 7 \\ -\ 7 \\ \hline \end{array}$$
$$\begin{array}{r} 18 \\ -\ 9 \\ \hline \end{array}$$
$$\begin{array}{r} 11 \\ -\ 3 \\ \hline \end{array}$$

2.
$$\begin{array}{r} 6 \\ -\ 0 \\ \hline \end{array}$$
$$\begin{array}{r} 14 \\ -\ 8 \\ \hline \end{array}$$
$$\begin{array}{r} 18 \\ -\ 9 \\ \hline \end{array}$$
$$\begin{array}{r} 13 \\ -\ 5 \\ \hline \end{array}$$
$$\begin{array}{r} 11 \\ -\ 8 \\ \hline \end{array}$$
$$\begin{array}{r} 14 \\ -\ 7 \\ \hline \end{array}$$
$$\begin{array}{r} 12 \\ -\ 3 \\ \hline \end{array}$$

3.
$$\begin{array}{r} 13 \\ -\ 9 \\ \hline \end{array}$$
$$\begin{array}{r} 15 \\ -\ 6 \\ \hline \end{array}$$
$$\begin{array}{r} 16 \\ -\ 7 \\ \hline \end{array}$$
$$\begin{array}{r} 12 \\ -\ 4 \\ \hline \end{array}$$
$$\begin{array}{r} 12 \\ -\ 8 \\ \hline \end{array}$$
$$\begin{array}{r} 8 \\ -\ 8 \\ \hline \end{array}$$
$$\begin{array}{r} 14 \\ -\ 5 \\ \hline \end{array}$$

4.
$$\begin{array}{r} 9 \\ -\ 0 \\ \hline \end{array}$$
$$\begin{array}{r} 11 \\ -\ 9 \\ \hline \end{array}$$
$$\begin{array}{r} 17 \\ -\ 9 \\ \hline \end{array}$$
$$\begin{array}{r} 14 \\ -\ 8 \\ \hline \end{array}$$
$$\begin{array}{r} 9 \\ -\ 9 \\ \hline \end{array}$$
$$\begin{array}{r} 12 \\ -\ 5 \\ \hline \end{array}$$
$$\begin{array}{r} 6 \\ -\ 6 \\ \hline \end{array}$$

5.
$$\begin{array}{r} 11 \\ -\ 4 \\ \hline \end{array}$$
$$\begin{array}{r} 7 \\ -\ 6 \\ \hline \end{array}$$
$$\begin{array}{r} 13 \\ -\ 8 \\ \hline \end{array}$$
$$\begin{array}{r} 14 \\ -\ 9 \\ \hline \end{array}$$
$$\begin{array}{r} 12 \\ -\ 9 \\ \hline \end{array}$$
$$\begin{array}{r} 8 \\ -\ 4 \\ \hline \end{array}$$
$$\begin{array}{r} 14 \\ -\ 9 \\ \hline \end{array}$$

6.
$$\begin{array}{r} 5 \\ -\ 5 \\ \hline \end{array}$$
$$\begin{array}{r} 0 \\ -\ 0 \\ \hline \end{array}$$
$$\begin{array}{r} 12 \\ -\ 6 \\ \hline \end{array}$$
$$\begin{array}{r} 14 \\ -\ 6 \\ \hline \end{array}$$
$$\begin{array}{r} 16 \\ -\ 8 \\ \hline \end{array}$$
$$\begin{array}{r} 7 \\ -\ 0 \\ \hline \end{array}$$
$$\begin{array}{r} 10 \\ -\ 5 \\ \hline \end{array}$$

7.
$$\begin{array}{r} 17 \\ -\ 8 \\ \hline \end{array}$$
$$\begin{array}{r} 15 \\ -\ 6 \\ \hline \end{array}$$
$$\begin{array}{r} 8 \\ -\ 0 \\ \hline \end{array}$$
$$\begin{array}{r} 9 \\ -\ 9 \\ \hline \end{array}$$
$$\begin{array}{r} 14 \\ -\ 7 \\ \hline \end{array}$$
$$\begin{array}{r} 11 \\ -\ 5 \\ \hline \end{array}$$
$$\begin{array}{r} 13 \\ -\ 4 \\ \hline \end{array}$$

Name _____

Learn

One fall day Mike picks 45 apples. He gives 31 apples to Mrs. Cole. How many apples does Mike have left?

Subtract to find how many are left.

 1 First subtract the ones.

tens	ones
4	5
− 3	1
	4

 2 Then subtract the tens.

tens	ones
4	5
− 3	1
1	4

____14____ apples are left.

You can use subtraction facts.

Try it   Subtract.

1.

tens	ones
4	7
− 3	5
1	2

tens	ones
5	2
− 2	1

tens	ones
7	5
−	2

tens	ones
8	6
− 4	3

2.

$$\begin{array}{r} 33 \\ -11 \\ \hline \end{array} \qquad \begin{array}{r} 64 \\ -42 \\ \hline \end{array} \qquad \begin{array}{r} 89 \\ -6 \\ \hline \end{array} \qquad \begin{array}{r} 63 \\ -61 \\ \hline \end{array} \qquad \begin{array}{r} 49 \\ -6 \\ \hline \end{array} \qquad \begin{array}{r} 32 \\ -11 \\ \hline \end{array}$$

3.

$$\begin{array}{r} 92 \\ -31 \\ \hline \end{array} \qquad \begin{array}{r} 75 \\ -3 \\ \hline \end{array} \qquad \begin{array}{r} 84 \\ -23 \\ \hline \end{array} \qquad \begin{array}{r} 58 \\ -4 \\ \hline \end{array} \qquad \begin{array}{r} 88 \\ -8 \\ \hline \end{array} \qquad \begin{array}{r} 49 \\ -2 \\ \hline \end{array}$$

 What subtraction facts can help you subtract 76 − 23?

Math at Home: Your child used basic facts to subtract tens and ones.
Activity: Have your child subtract 48 - 23.

Practice Subtract.

Subtract the ones first.

4.

tens	ones
5	8
− 2	3
3	5

tens	ones
7	6
− 7	4

tens	ones
9	7
− 3	4

tens	ones
8	5
− 1	1

5.

$$\begin{array}{r} 64 \\ -\ 21 \\ \hline \end{array}$$
$$\begin{array}{r} 78 \\ -\ 65 \\ \hline \end{array}$$
$$\begin{array}{r} 77 \\ -\ 3 \\ \hline \end{array}$$
$$\begin{array}{r} 65 \\ -\ 11 \\ \hline \end{array}$$
$$\begin{array}{r} 59 \\ -\ 6 \\ \hline \end{array}$$
$$\begin{array}{r} 67 \\ -\ 42 \\ \hline \end{array}$$

6.

$$\begin{array}{r} 74 \\ -\ 43 \\ \hline \end{array}$$
$$\begin{array}{r} 89 \\ -\ 15 \\ \hline \end{array}$$
$$\begin{array}{r} 65 \\ -\ 62 \\ \hline \end{array}$$
$$\begin{array}{r} 59 \\ -\ 6 \\ \hline \end{array}$$
$$\begin{array}{r} 86 \\ -\ 41 \\ \hline \end{array}$$
$$\begin{array}{r} 48 \\ -\ 42 \\ \hline \end{array}$$

7.

$$\begin{array}{r} 49 \\ -\ 4 \\ \hline \end{array}$$
$$\begin{array}{r} 36 \\ -\ 2 \\ \hline \end{array}$$
$$\begin{array}{r} 85 \\ -\ 53 \\ \hline \end{array}$$
$$\begin{array}{r} 74 \\ -\ 31 \\ \hline \end{array}$$
$$\begin{array}{r} 98 \\ -\ 2 \\ \hline \end{array}$$
$$\begin{array}{r} 57 \\ -\ 14 \\ \hline \end{array}$$

8.

$$\begin{array}{r} 81 \\ -\ 1 \\ \hline \end{array}$$
$$\begin{array}{r} 43 \\ -\ 22 \\ \hline \end{array}$$
$$\begin{array}{r} 12 \\ -\ 7 \\ \hline \end{array}$$
$$\begin{array}{r} 48 \\ -\ 11 \\ \hline \end{array}$$
$$\begin{array}{r} 54 \\ -\ 10 \\ \hline \end{array}$$
$$\begin{array}{r} 36 \\ -\ 32 \\ \hline \end{array}$$

9.

$$\begin{array}{r} 78 \\ -\ 42 \\ \hline \end{array}$$
$$\begin{array}{r} 56 \\ -\ 16 \\ \hline \end{array}$$
$$\begin{array}{r} 33 \\ -\ 22 \\ \hline \end{array}$$
$$\begin{array}{r} 95 \\ -\ 94 \\ \hline \end{array}$$
$$\begin{array}{r} 60 \\ -\ 10 \\ \hline \end{array}$$
$$\begin{array}{r} 25 \\ -\ 5 \\ \hline \end{array}$$

What could the next number be? Find the pattern.

10. 74 64 54 44 ☐   47 37 27 17 ☐

Name _____

Learn

You may need to regroup.

Use tens and ones models to subtract 34 − 8.

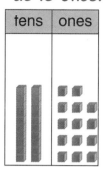 **1** Show 34.

tens	ones

3 tens 4 ones

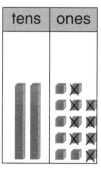 **2** Regroup 1 ten as 10 ones.

tens	ones

2 tens 14 ones

3 Subtract 8 ones.

tens	ones

2 tens 6 ones are left.

$$34 - 8 = \underline{26}$$

Try it

Use ▪ and ▬▬▬▬ .

Subtract.	Do you need to regroup?	How many are left?
1. 42 − 7	(yes) no	35
2. 63 − 5	yes no	___
3. 28 − 5	yes no	___
4. 51 − 6	yes no	___

Workspace

Sum it Up! How do you know if you need to regroup to find 63 − 45?

 **Math at Home:** Your child subtracted two numbers, deciding each time if regrouping was needed.
Activity: Ask your child why regrouping is necessary to subtract 46 - 28.

two hundred seven **207**

Use ▢ and ▭ .

Regroup when you need to.

	Subtract.	Do you need to regroup?	How many are left?
5.	53 − 8	(yes) no	45
6.	44 − 9	yes no	___
7.	74 − 3	yes no	___
8.	93 − 7	yes no	___
9.	85 − 3	yes no	___
10.	32 − 6	yes no	___
11.	26 − 3	yes no	___
12.	52 − 3	yes no	___

Workspace

Problem Solving

Number Sense

13. What numbers between 1 and 9 can you subtract from 75 without regrouping?

14. What numbers between 1 and 9 can you subtract from 93 without regrouping?

Learn

Subtract 42 − 15.

 Are there enough ones to subtract 5 ones?

Regroup if you need to.

1 Show 42.

tens	ones
4	2
− 1	5

2 Regroup 1 ten as 10 ones if you need to.

tens	ones
③ 4	⑫ 2
− 1	5

3 Subtract the ones.

tens	ones
③ 4	⑫ 2
− 1	5
	7

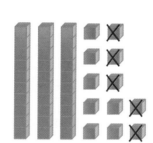

4 Subtract the tens.

tens	ones
③ 4	⑫ 2
− 1	5
2	7

Try it   Subtract.

1.

tens	ones
② 3	⑭ 4
− 1	7
1	7

tens	ones
6	3
−	5

tens	ones
5	7
− 1	2

tens	ones
2	5
− 1	8

 **Sum it Up** How do you show regrouping 1 ten as 10 ones in 48 − 19?

 Math at Home: Your child subtracted, regrouping 1 ten as 10 ones when needed.
Activity: Ask your child to show you how to subtract 43 - 17.

Practice Subtract.

Decide if you need to regroup.

2.

tens	ones
⁶⁄7̶	¹³ 3̶
− 4	8
2	5

tens	ones
☐	☐
6	1
− 5	7

tens	ones
☐	☐
4	2
−	1

tens	ones
☐	☐
5	5
− 4	6

3.

tens	ones
☐	☐
3	4
− 2	6

tens	ones
☐	☐
4	6
− 2	4

tens	ones
☐	☐
5	3
− 1	4

tens	ones
☐	☐
6	2
−	8

4.

tens	ones
☐	☐
5	2
−	8

tens	ones
☐	☐
3	9
− 1	2

tens	ones
☐	☐
8	3
− 2	7

tens	ones
☐	☐
7	6
− 4	6

Spiral Review and Test Prep

Choose the correct answer.

5. 77 − 48

- ◯ 29
- ◯ 31
- ◯ 39
- ◯ 41

6. 32 + 27

- ◯ 29
- ◯ 39
- ◯ 49
- ◯ 59

Name _____

Learn

Subtract 51 – 6 and 46 – 34.

Do you need to regroup?

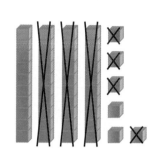

There are not enough ones.
Regroup 1 ten as 10 ones.

tens	ones
⁴5̷	¹¹1̷
– 2	6
4	5

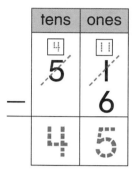

There are enough ones to subtract.

tens	ones
□ 4	□ 6
– 3	4
1	2

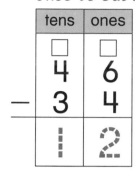

Try it

Subtract. You can use tens and ones models.
Regroup if you need to.

1.

tens	ones
³4̷	¹⁵5̷
– 2	6
1	9

tens	ones
□ 5	□ 2
–	5

tens	ones
□ 6	□ 8
– 2	4

tens	ones
□ 3	□ 5
– 2	8

2.

tens	ones
□ 7	□ 2
– 1	5

tens	ones
□ 3	□ 4
–	9

tens	ones
□ 5	□ 6
– 3	4

tens	ones
□ 4	□ 1
– 2	8

Sum it Up

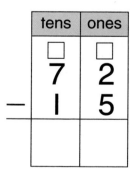

How do you know when you need to regroup?

Math at Home: Your child decided whether to regroup, then subtracted.
Activity: Ask your child to show you subtraction that requires regrouping and
subtraction that does not require regrouping.

Practice

Subtract. You can use tens and ones models.

Decide if you need to regroup.

3.

tens	ones
7̶ 8	1̶2̶ 2̶
− 5	7
2	5

tens	ones
☐ 8	☐ 9
− 3	7

tens	ones
☐ 4	☐ 1
−	6

tens	ones
☐ 6	☐ 4
− 3	7

4.

tens	ones
☐ 5	☐ 1
− 2	8

tens	ones
☐ 4	☐ 2
− 1	8

tens	ones
☐ 5	☐ 7
− 1	4

tens	ones
☐ 7	☐ 6
−	8

5.

tens	ones
☐ 3	☐ 8
−	9

tens	ones
☐ 3	☐ 7
− 1	5

tens	ones
☐ 3	☐ 9
− 2	3

tens	ones
☐ 5	☐ 6
− 4	7

Problem Solving

Data

6. How many more sunny days were there in June than in July?

_____ days

7. How many more sunny days were there in August than in July?

_____ days

☀ SUNNY DAYS ☀	
JUNE	23 days
JULY	15 days
AUGUST	20 days

Learn

Carley picked 30 strawberries from the garden. She put 23 of them in a bowl.
She ate the rest.
How many did she eat?

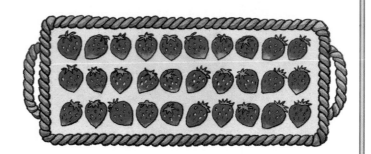

1 Regroup if you need to.

tens	ones
2 3	10 0
– 2	3

2 Subtract the ones. Then subtract the tens.

tens	ones
2 3̸	10 0̸
– 2	3
	7

 Do you need to regroup?

_7_ strawberries

Try it

Subtract.
You can use tens and ones models.

1.

```
 3 8      2 7      4 5      3 2      6 2      5 6
–   6    – 1 8    – 2 3    –   6    – 2 7    – 5 4
```

32

  How do you subtract 2-digit numbers?

 **Math at Home:** Your child practiced 2-digit subtraction with and without regrouping.
Activity: Ask your child to show you how to subtract 46 – 28.

two hundred thirteen **213**

McGraw-Hill School Division

Practice

Subtract. You can use tens
and ones models.

Regroup if
you need to.

2.

□	□ □
2 1	
− 1 5	
6	

□	□
3 8	
− 1 2	

□	□
4 6	
− 5	

□	□
5 4	
− 3 8	

□	□
7 2	
− 6 9	

□	□
3 9	
− 2 2	

3.

□	□
4 2	
− 2 7	

□	□
6 2	
− 5 3	

□	□
3 6	
− 7	

□	□
1 9	
− 2	

□	□
3 4	
− 1 6	

□	□
6 7	
− 5 2	

4.

□	□
3 0	
− 1 8	

□	□
5 2	
− 9	

□	□
7 8	
− 2 4	

□	□
4 2	
− 1 8	

□	□
6 1	
− 5 8	

□	□
9 8	
− 4 5	

5.

□	□
5 8	
− 2 0	

□	□
6 3	
− 5 9	

□	□
7 2	
− 8	

□	□
8 4	
− 4	

□	□
4 1	
− 1 0	

□	□
3 3	
− 2 5	

Rewrite. Then subtract.

6. 87 − 25

$$
\begin{array}{r}
87 \\
- 25 \\
\hline
62
\end{array}
$$

 52 − 28

+ _____

 52 − 30

+ _____

Name _____

Steps in a Process

Reading Skill   You can use steps to help you solve problems.

Mr. Brown's class is making a garden. First, they paint signs for the plants. Next they plant 40 pansy seeds. Then they plant 32 tomato plants. Finally, they plant 10 sunflowers.

Answer each question.

1. What do the children do **first**? _____

2. What do they do **next**? _____

3. What do they do **after that**? _____

4. What do the children **finally** do? _____

5. Write a subtraction problem about the story.

_____

Math at Home: Your child read a story and named steps in a process.
Activity: Have your child name steps to make a favorite snack.

McGraw-Hill School Division

Answer each question.

Zeke and Ben work in the garden on Friday. First, they pull up weeds. Next, they water all the plants. After that, they pick 12 tomatoes. Finally, they pick 8 flowers.

6. What do Zeke and Ben do **first**? _____

7. What do they do **next**? _____

8. What do they do **after that**? _____

9. What do Luke and Ben do **last**? _____

10. Write a subtraction problem about the story.

_____

Tell how following the steps in a process can help you solve problems.

_____

Check Your Progress A

Name _____

Subtract.

1. 60 − 40 = _____   82 − 30 = _____   58 − 20 = _____

2.
$$\begin{array}{r} 47 \\ -\ 15 \\ \hline \end{array} \qquad \begin{array}{r} 22 \\ -\ 10 \\ \hline \end{array} \qquad \begin{array}{r} 51 \\ -21 \\ \hline \end{array} \qquad \begin{array}{r} 68 \\ -\ 27 \\ \hline \end{array} \qquad \begin{array}{r} 76 \\ -\ 20 \\ \hline \end{array}$$

Subtract. Regroup if you need to.

3.

tens	ones
□ 5	□ 2
− 2	6

tens	ones
□ 6	□ 4
− 3	9

tens	ones
□ 8	□ 3
− 4	7

tens	ones
□ 7	□ 8
−	3

4.

tens	ones
□ 3	□ 5
− 2	3

tens	ones
□ 6	□ 2
− 2	3

tens	ones
□ 4	□ 2
−	8

tens	ones
□ 8	□ 6
− 4	6

5.

tens	ones
□ 2	□ 1
− 1	7

tens	ones
□ 8	□ 3
−	7

tens	ones
□ 5	□ 6
− 3	8

tens	ones
□ 4	□ 5
−	8

Complete each fact family.

1.
$6 + 9 = $ _____ $4 + 5 = $ _____   _____ $ + 7 = 10$

$9 + 6 = $ _____ $5 + 4 = $ _____ $7 \bigcirc 3 = 10$

$15 - $ _____ $= 9$ $9 \bigcirc 4 = 5$ _____ $- 7 = 3$

_____ $- 9 = 6$ $9 - $ _____ $= 4$ $10 - $ _____ $= 7$

Write the number that comes just after.

2. 27 28 _____   58 59 _____   92 93 _____

Write each number as tens and ones and then in expanded form.

3. $68 = $ _____ tens _____ ones $= $ _____ $+$ _____

4. $53 = $ _____ tens _____ ones $= $ _____ $+$ _____

TECHNOLOGY LINK

Use Place Value Models to Subtract
- Choose place value.
- Choose a mat to subtract.
- Stamp out 53. Click on $-$.
- Trade one 10 down.
- Click on 16.
- What is the difference? _____

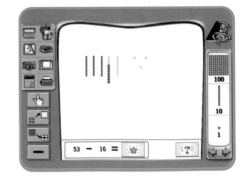

  1. Choose place value. Show 42 - 24.
  What is the difference? _____

  2. Stamp out another number. Choose a number
  to subtract. Tell the difference.

For more practice use Math Traveler™.

Name _____

Choose the Operation

Read ➤ The second graders pick 32 pumpkins. The first graders pick 28 pumpkins. How many more pumpkins do the second graders pick than the first graders?

What do I know?

What do I need to find out?

Plan ➤ Should I add or subtract? The question asks how many more, so I can subtract.

Solve ➤ $32 - 17 = 15$
The second graders pick 15 more pumpkins than the first graders.

Look Back ➤ Does my answer make sense? Why?

Circle add or subtract. Solve.

1. Crissy collects 24 orange leaves and 8 yellow leaves. How many more orange leaves does she collect than yellow leaves?

   add   subtract

   _____

   _____ orange leaves

2. Alex picks 30 green apples. Deidra picks 20 red apples. How many apples do they pick in all?

   add   subtract

   _____

   _____ apples

 How do you know when to subtract to solve a problem?

 Math at Home: Your child learned to choose the operation needed to solve problems.
Activity: Ask your child to tell a math story that can be solved by subtracting.

two hundred nineteen **219**

Circle add or subtract. Solve.

3. There are 15 boys and 16 girls working on the winter mural. How many children are working in all?

add   subtract

_____

_____ children

4. Last fall the children planted 31 tulip bulbs. All but 3 bulbs came up in the spring. How many tulip bulbs came up?

add   subtract

_____

_____ tulip bulbs

5. There are 19 girls raking leaves. 3 boys join them. How many children are raking leaves?

add   subtract

_____

_____ children

6. 23 children in Mr. Luke's class say summer is their favorite season. 8 children say winter. How many more children like summer?

add   subtract

_____

_____ students

7. Look at the picture. Write a problem where you use both addition and subtraction to solve.

_____

_____

Learn

The children at Blaine School
built 21 bird feeders.
All but 3 were sold at the school fair.
How many bird feeders were sold?

You can add to check subtraction.

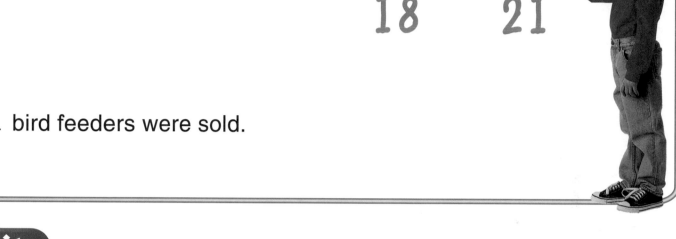

$$
\begin{array}{r}
\overset{1\,11}{2\!\!\!/1} \\
-\ 3 \\
\hline
18
\end{array}
\qquad
\begin{array}{r}
\overset{1}{18} \\
+\ 3 \\
\hline
21
\end{array}
$$

__18__ bird feeders were sold.

Try it   Subtract. Check by adding.

1.

$$
\begin{array}{r}
52 \\
-\ 31 \\
\hline
21
\end{array}
\qquad
+
\begin{array}{r}
21 \\
31 \\
\hline
52
\end{array}
$$

$$
\begin{array}{r}
54 \\
-\ 35 \\
\hline
\end{array}
\qquad +
$$

$$
\begin{array}{r}
49 \\
-\ 16 \\
\hline
\end{array}
\qquad +
$$

2.

$$
\begin{array}{r}
32 \\
-\ 17 \\
\hline
\end{array}
\qquad +
$$

$$
\begin{array}{r}
63 \\
-\ 9 \\
\hline
\end{array}
\qquad +
$$

$$
\begin{array}{r}
85 \\
-\ 36 \\
\hline
\end{array}
\qquad +
$$

Sum it Up   How can you add to check $27 - 8 = 19$?

Math at Home: Your child learned to check subtraction by adding.
Activity: Ask your child to subtract 45 - 29. Have your child find the difference and
then add to check the answer.

Subtract. Check by adding.

To check, start with the difference.

3.

$$70 - 27 = 43$$ $$43 + 27 = 70$$

$$73 - 55 = \underline{\hspace{1cm}} + \underline{\hspace{1cm}}$$

$$56 - 29 = \underline{\hspace{1cm}} + \underline{\hspace{1cm}}$$

4.

$$74 - 43 = \underline{\hspace{1cm}} + \underline{\hspace{1cm}}$$

$$62 - 57 = \underline{\hspace{1cm}} + \underline{\hspace{1cm}}$$

$$26 - 8 = \underline{\hspace{1cm}} + \underline{\hspace{1cm}}$$

5.

$$42 - 38 = \underline{\hspace{1cm}} + \underline{\hspace{1cm}}$$

$$56 - 32 = \underline{\hspace{1cm}} + \underline{\hspace{1cm}}$$

$$73 - 68 = \underline{\hspace{1cm}} + \underline{\hspace{1cm}}$$

6.

$$37 - 14 = \underline{\hspace{1cm}} + \underline{\hspace{1cm}}$$

$$81 - 27 = \underline{\hspace{1cm}} + \underline{\hspace{1cm}}$$

$$65 - 37 = \underline{\hspace{1cm}} + \underline{\hspace{1cm}}$$

7.

$$66 - 8 = \underline{\hspace{1cm}} + \underline{\hspace{1cm}}$$

$$48 - 19 = \underline{\hspace{1cm}} + \underline{\hspace{1cm}}$$

$$50 - 27 = \underline{\hspace{1cm}} + \underline{\hspace{1cm}}$$

Workspace

8. Draw a picture. Show the difference of 33 − 10 using tens and ones.

Learn

Every spring many butterflies fly from Mexico to California. Tracy counts 42 butterflies. Kendra counts 28 butterflies. How many more butterflies does Tracy count than Kendra?

$$42 - 28 = 14$$

Tracy counts 14 more butterflies than Kendra.

You can estimate to see if the answer is reasonable.

| 10 | 20 | 28 30 | 40 42 | 50 | 60 |

$$\begin{array}{r} 42 \\ -\ 28 \\ \hline 14 \end{array}$$ nearest tens → nearest tens → $$\begin{array}{r} 40 \\ -\ 30 \\ \hline 10 \end{array}$$

Rewrite the numbers to the nearest 10.

14 is close to 10.
The answer is reasonable.

Try it

Subtract. Estimate to see if your answer is reasonable.

1.

$$\begin{array}{r} 58 \\ -\ 21 \\ \hline 37 \end{array}$$ — $$\begin{array}{r} 60 \\ -\ 20 \\ \hline 40 \end{array}$$

$$\begin{array}{r} 49 \\ -\ 37 \\ \hline \end{array}$$ —

$$\begin{array}{r} 31 \\ -\ 22 \\ \hline \end{array}$$ —

Sum it Up

How can you use estimation to check subtraction?

Math at Home: Your child estimated differences by using the nearest tens.
Activity: Ask your child to subtract 38 - 27. Then, have him or her explain how to estimate to check the answer.

Subtract. Estimate to see if your answer is reasonable.

Remember to use the nearest tens.

2.

$$\begin{array}{r} 62 \\ -\ 18 \\ \hline 44 \end{array} \qquad \begin{array}{r} 60 \\ -\ 20 \\ \hline 40 \end{array}$$

$$\begin{array}{r} 59 \\ -\ 33 \\ \hline \end{array} \qquad -\ \_\_$$

$$\begin{array}{r} 46 \\ -\ 27 \\ \hline \end{array} \qquad -\ \_\_$$

3.

$$\begin{array}{r} 88 \\ -\ 46 \\ \hline \end{array} \qquad -\ \_\_$$

$$\begin{array}{r} 31 \\ -\ 12 \\ \hline \end{array} \qquad -\ \_\_$$

$$\begin{array}{r} 94 \\ -\ 54 \\ \hline \end{array} \qquad -\ \_\_$$

4.

$$\begin{array}{r} 53 \\ -\ 29 \\ \hline \end{array} \qquad -\ \_\_$$

$$\begin{array}{r} 77 \\ -\ 48 \\ \hline \end{array} \qquad -\ \_\_$$

$$\begin{array}{r} 67 \\ -\ 41 \\ \hline \end{array} \qquad -\ \_\_$$

5.

$$\begin{array}{r} 87 \\ -\ 31 \\ \hline \end{array} \qquad -\ \_\_$$

$$\begin{array}{r} 28 \\ -\ 19 \\ \hline \end{array} \qquad -\ \_\_$$

$$\begin{array}{r} 72 \\ -\ 28 \\ \hline \end{array} \qquad -\ \_\_$$

 Problem Solving

Patterns

Look at each shape pattern.
What could the next shape be?

6. 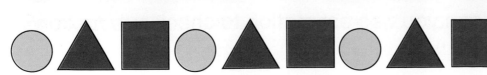 ____

7. ____

Name _____

Learn

Jimmy buys a pumpkin for 59¢.
He gives the clerk 75¢.
How much change does he get back?

$$\begin{array}{r} 75¢ \\ - 59¢ \\ \hline 16¢ \end{array}$$

Jimmy gets 16¢ change.

Pumpkin 59¢

Try it   Subtract.

1.
$$\begin{array}{r} 42¢ \\ - 15¢ \\ \hline 27¢ \end{array}$$
$$\begin{array}{r} 60¢ \\ - 45¢ \\ \hline \end{array}$$
$$\begin{array}{r} 75¢ \\ - 68¢ \\ \hline \end{array}$$
$$\begin{array}{r} 58¢ \\ - 26¢ \\ \hline \end{array}$$
$$\begin{array}{r} 76¢ \\ - 27¢ \\ \hline \end{array}$$
$$\begin{array}{r} 89¢ \\ - 25¢ \\ \hline \end{array}$$

2.
$$\begin{array}{r} 50¢ \\ - 10¢ \\ \hline \end{array}$$
$$\begin{array}{r} 78¢ \\ - 50¢ \\ \hline \end{array}$$
$$\begin{array}{r} 22¢ \\ - 8¢ \\ \hline \end{array}$$
$$\begin{array}{r} 34¢ \\ - 17¢ \\ \hline \end{array}$$
$$\begin{array}{r} 50¢ \\ - 25¢ \\ \hline \end{array}$$
$$\begin{array}{r} 62¢ \\ - 58¢ \\ \hline \end{array}$$

3.
$$\begin{array}{r} 60¢ \\ - 32¢ \\ \hline \end{array}$$
$$\begin{array}{r} 65¢ \\ - 46¢ \\ \hline \end{array}$$
$$\begin{array}{r} 51¢ \\ - 18¢ \\ \hline \end{array}$$
$$\begin{array}{r} 86¢ \\ - 2¢ \\ \hline \end{array}$$
$$\begin{array}{r} 63¢ \\ - 27¢ \\ \hline \end{array}$$
$$\begin{array}{r} 90¢ \\ - 34¢ \\ \hline \end{array}$$

How is subtracting money amounts like subtracting other numbers?

Math at Home: Your child subtracted amounts of money up to 99¢.
Activity: Have your child show you how to subtract 50¢ - 35¢.

Practice

Subtract.

Remember to write ¢.

4.

59¢	40¢	76¢	61¢	36¢	29¢
− 34¢	− 7¢	− 32¢	− 17¢	− 29¢	− 8¢
25¢					

5.

17¢	48¢	73¢	23¢	60¢	88¢
− 5¢	− 22¢	− 48¢	− 18¢	− 40¢	− 53¢

6.

98¢	34¢	12¢	40¢	53¢	33¢
− 42¢	− 19¢	− 7¢	− 38¢	− 27¢	− 15¢

7.

51¢	80¢	43¢	77¢	25¢	45¢
− 29¢	− 64¢	− 15¢	− 23¢	− 9¢	− 25¢

Problem Solving

8. Mindy buys a pumpkin for 66¢. She gives the clerk 75¢. She gets back 12¢ change. Is this correct? Why or why not?

_____

9. Ben buys a little pumpkin for 25¢.
He gives the clerk 50¢.
How much change does he receive?

_____

Name

Plant a Garden

You want to plant a garden.
Here are seeds you can choose
from to buy. You have 99¢.

Plan which seeds you want to buy.
Make a list.

1. _____

Workspace

Problem Solving • Decision Making

Practice.

2. How much did you spend?
Show how you found out.

_____

_____

_____

_____

3. What was your change?
Show how you found out.

_____

_____

_____

_____

4. What if you only had 75¢?
What seed choices would
you have?

_____

_____

_____

_____

Name _____

Why Is Weather Important?

You Will Need

thermometer

What to do

1. Observe the weather for 5 days.

2. Observe in the morning.

3. Fill in the weather chart. Use the weather symbols. Record the temperature.

4. Repeat for the afternoon.

Weather symbols

sunny   cloudy   rain   snow   windy

		Monday	Tuesday	Wednesday	Thursday	Friday
Morning	Weather					
	Temperature					
Afternoon	Weather					
	Temperature					

What did you find out?

1. How did the weather change each day?

_____

2. On Monday, was the afternoon temperature higher or lower than the morning temperature?

_____

Write a problem to show the difference.

_____

3. Did you see any patterns during the week?

_____

 Want to do more?
Observe the weather for another week.
What patterns did you see? Explain.

Name _____

Subtract. Check by adding.

1.
```
    42              73              96
  − 25    + __    − 38    + __    − 24    + __
```

Subtract. Estimate to see if your answer is reasonable.

2.
```
    60              78              36
  − 36    − __    − 22    − __    − 27    − __
```

Subtract.

3.
```
   45¢      86¢      75¢      37¢      51¢      32¢
 − 17¢    − 40¢    −  9¢    − 12¢    − 32¢    − 27¢
```

4.
```
   50¢      68¢      34¢      57¢      82¢      41¢
 − 18¢    − 26¢    − 25¢    −  9¢    − 27¢    − 16¢
```

Circle add or subtract. Solve.

5. The children picked 26 acorn squash. They also picked 7 yellow butternut squash. How many more acorn squash did they pick?

    add   subtract   − _____

    _____ acorn squash

6. Libby picked 54 apples. George picked 37 apples. How many apples did they pick in all?

    add   subtract   + _____

    _____ apples

MATH GAME

Name _____

Bird Calls

Take turns.

- Choose a crayon. Choose a bird. Subtract.

- Have your partner check your work.

- If correct, color the bird with your crayon.

- The player who colors more birds is the winner.

$$30 - 20$$

$$62 - 41$$

$$84 - 4$$

$$46 - 38$$

$$25 - 9$$

$$60 - 10$$

$$76 - 26$$

$$93 - 8$$

$$51 - 6$$

$$38 - 28$$

$$52 - 45$$

$$79 - 8$$

$$94 - 37$$

$$46 - 11$$

$$82 - 46$$

$$35 - 23$$

Read these words.

Name _____

Math Words

subtract
regroup
tens
ones

Language and Math

Complete. Use a word from the list.

1. You have to _____ when you subtract 46 − 28.

2. The number 54 has 5 _____.

3. You _____ to find the difference of 56 − 23.

Concepts and Skills

Subtract.

4.

62	38	45	73	87	56
− 35	− 18	− 27	− 49	− 32	− 8

5.

78	69	43	26	55¢	76¢
− 44	− 51	− 29	− 7	− 18¢	− 30¢

6.

32	86	60	64	45¢	91¢
− 16	− 81	− 20	− 42	− 9¢	− 27¢

Subtract.

7.
$$36¢ - 21¢$$
$$55 - 23$$
$$60¢ - 27¢$$
$$47 - 9$$
$$80 - 30$$

Subtract. Check by adding.

8.
$$47 - 23$$
$$42 - 28$$
$$78 - 52$$
$$24 - 14$$
$$63¢ - 55¢$$

Problem Solving

Circle add or subtract. Solve.

9. There are 24 boys and 36 girls sledding. How many children are there altogether?

add   subtract

$$+ \underline{\qquad}$$

_____ children

10. There are 23 children skating. 9 of them are leaving. How many children are still skating?

add   subtract

$$- \underline{\qquad}$$

_____ children

Journal

How can you tell when to add and when to subtract? Explain.

_____

Name _____

Subtract. Then color the butterflies.

Differences between 40 and 50:  yellow

Differences between 50 and 60: 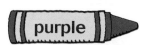 purple

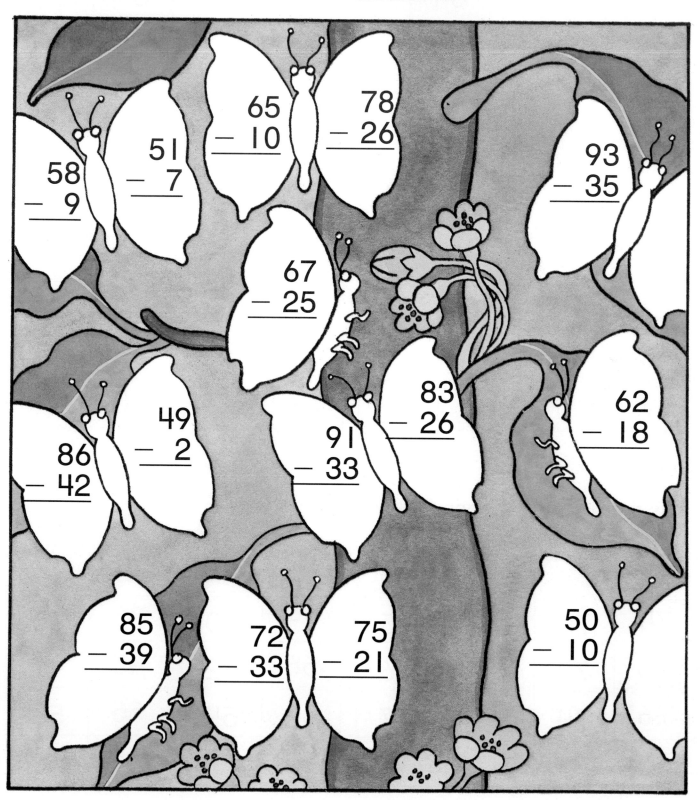

Algebra & functions

Look at the numbers. What patterns do you see?

60
62
64
66

Subtract 10. What patterns do you see?

	Subtract 10
60	50
62	52
64	54
66	56

Complete each table.

1.

	Subtract 2
56	
57	
58	
59	

	Subtract 7
74	
76	
78	
80	

	Subtract 9
82	
84	
86	
88	

	Subtract 10
55	
60	
65	
70	

What is the pattern for each table?

2.

	Subtract ___
74	72
75	73
76	74
77	75

	Subtract ___
88	85
89	86
90	87
91	88

	Subtract ___
54	50
56	52
58	54
60	56

	Subtract ___
35	30
45	40
55	50
65	60

Name _____

Subtract.

1.
$$
\begin{array}{r} 58 \\ -\ 34 \\ \hline \end{array}
\qquad
\begin{array}{r} 50 \\ -\ 20 \\ \hline \end{array}
\qquad
\begin{array}{r} 53 \\ -\ 25 \\ \hline \end{array}
\qquad
\begin{array}{r} 76 \\ -\ \ 2 \\ \hline \end{array}
\qquad
\begin{array}{r} 32 \\ -\ 18 \\ \hline \end{array}
\qquad
\begin{array}{r} 62 \\ -\ 37 \\ \hline \end{array}
$$

2.
$$
\begin{array}{r} 41 \\ -\ 39 \\ \hline \end{array}
\qquad
\begin{array}{r} 32 \\ -\ 15 \\ \hline \end{array}
\qquad
\begin{array}{r} 78¢ \\ -\ 42¢ \\ \hline \end{array}
\qquad
\begin{array}{r} 39 \\ -\ 23 \\ \hline \end{array}
\qquad
\begin{array}{r} 56¢ \\ -\ 30¢ \\ \hline \end{array}
\qquad
\begin{array}{r} 32 \\ -\ 24 \\ \hline \end{array}
$$

3.
$$
\begin{array}{r} 42 \\ -\ 27 \\ \hline \end{array}
\qquad
\begin{array}{r} 75¢ \\ -\ 49¢ \\ \hline \end{array}
\qquad
\begin{array}{r} 45 \\ -\ 19 \\ \hline \end{array}
\qquad
\begin{array}{r} 81 \\ -\ 76 \\ \hline \end{array}
\qquad
\begin{array}{r} 98 \\ -\ \ 6 \\ \hline \end{array}
\qquad
\begin{array}{r} 60 \\ -\ 25 \\ \hline \end{array}
$$

Circle add or subtract. Solve.

4. There are 35 children at the beach.
   7 leave. How many children
   are still at the beach? _____ children   add   subtract

5. In September, there were 21
   sunny days. October had 12
   sunny days. How many more
   sunny days were there in
   September than in October? _____ days   add   subtract

Spring Flowers

The second grade class took a survey to find their favorite flower. Use the clues to tell how the class voted.

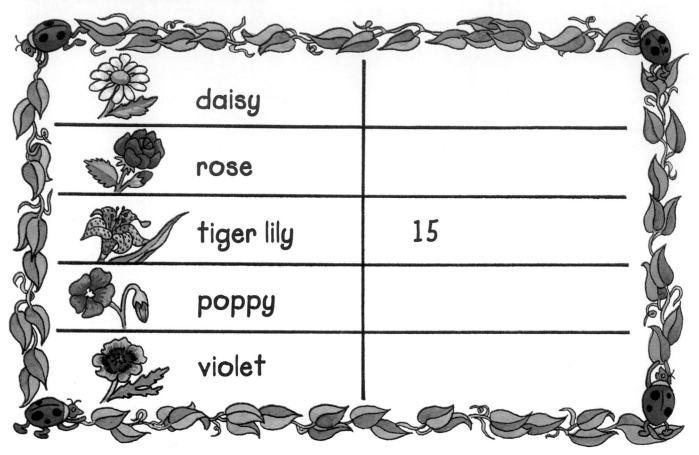

daisy		
rose		
tiger lily	15	
poppy		
violet		

1. The daisy got 26 more votes than the violet.

2. The rose got 27 votes fewer than the poppy.

3. The tiger lily got 17 votes fewer than the poppy.

4. The violet got 5 fewer votes than the tiger lily.

5. Explain how you decided where to place each flower in the survey.

_____

6. Which flower is the class favorite? _____

You may want to add this page to your portfolio.

Name _____

Number Sense

Choose the correct answer.

1. Which statement is true?

- ◯ 83 < 92
- ◯ 83 = 92
- ◯ 83 > 92
- ◯ 92 = 83

2. There are 42 children playing. 8 leave. How many children are still playing?

- ◯ 8
- ◯ 34
- ◯ 42
- ◯ 50

3. What number comes after 63?

62	63	

- ◯ 62
- ◯ 63
- ◯ 64
- ◯ 65

Algebra and Functions

4. Which number sentence can you use to solve this problem?

Leon planted 3 trees. Jo planted 4 trees. How many trees did they plant altogether?

- ◯ $3 + 4 = 7$
- ◯ $7 - 3 = 4$
- ◯ $4 - 3 = 1$
- ◯ $7 - 4 = 3$

5. What goes in the box?

$8 \boxed{} 5 = 13$

- ◯ =
- ◯ +
- ◯ −
- ◯ ×

6. There are 7 birds. Which shows how many birds are left in the tree?

- ◯ $7 + 5 = 1$
- ◯ $7 - 5 = 2$
- ◯ $5 + 7 = 1$
- ◯ $7 - 2 = 5$

Mathematical Reasoning

Choose the correct answer.

7. I am a number between 40 and 43. When you count by 2, you say my name. What number am I?

- ◯ 40
- ◯ 41
- ◯ 42
- ◯ 43

8. Sara says the answer to this problem is 6. Is she correct? Tell how you know.

Sean has 12 stamps. He gives 6 stamps to his sister. How many does he have left?

_____

_____

9. Alexa has $3.15. Her mother gives her 2 quarters. How much does she have now?

- ◯ $3.15
- ◯ $3.65
- ◯ $3.90
- ◯ $3.95

Statistics, Data Analysis and Probability

10. What could the next shape be in the shape pattern?

■▲●◯■▲●◯■▲●◯■

- ◯ ■
- ◯ ●
- ◯ ▲
- ◯ ▬

Use the graph for 11 and 12.

Books Read					
Lena	▯	▯			
Brent	▯	▯	▯	▯	
James	▯	▯	▯	▯	▯
Tammy	▯	▯	▯		

Each ▯ stands for 1 book.

11. Who read 6 books?

- ◯ Lena
- ◯ Brent
- ◯ James
- ◯ Tammy

12. Who read the least?

- ◯ Lena
- ◯ Brent
- ◯ James
- ◯ Tammy

WALLY'S PUPPET SHOW

Next Show at 3:00!

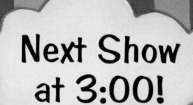

theme
It's Show Time!

Use the Data
What time does the clock show?

What You Will Learn
In this chapter you will learn how to:

- Tell time.
- Find elapsed time.
- Read a calendar.
- Use act it out to solve problems about time.

Let the Show Begin

Story by Susan Freeman   Illustrated by Deborah Melmon

MATH AT HOME

Dear Family,

In Chapter 7, I will learn to tell time and read a calendar. Here are new vocabulary words and an activity that we can do together.

My Daily Schedule

- Ask your child what time he or she got up this morning. Record the time.

- Ask your child to set the hands on a clock to show the time.

- Repeat using times of other daily events, such as the times school starts and ends, dinnertime, and bedtime.

use

clock

paper and pencil

Math Words

hour

60 minutes

  8:00

minute

60 seconds

 minute

half hour

30 minutes

calendar

January						
S	M	T	W	T	F	S
		1	2	3	4	5
6	7	8	9	10	11	12
13	14	15	16	17	18	19
20	21	22	23	24	25	26
27	28	29	30	31		

Additional activities at
www.mhschool.com/math

Name _____

Learn

What time does each clock show?

hour hand

minute hand

8:00

eight o'clock

8:30

eight-thirty

Math Words

hour
half hour
minute
hour hand
minute hand

One hour is 60 minutes.

One minute is 60 seconds.

One half hour is 30 minutes.

Try it

Write each time.

1.

6:00

2.

Sum it Up

Where does the minute hand point at 3:30?

 Math at Home: Your child practiced telling time to the hour and half hour.
Activity: Have your child show you where the clock hands are for 8:00.
Repeat using different times.

Practice Write each time.

3.

 10:30

 :

 :

 :

4.

 :

 :

 :

 :

Draw the minute hand to show each time.

5.

 8:30

 1:00

6.

 3:00

 7:30

Learn

The puppet show begins at 7:00.
At 7:25 the sock hoppers dance.

You can count
by fives to tell
the time.

7:00

7:25

Try it   Write each time.

1.

6:00   :   :   :

2.

:   :   :   :

Sum it Up

How many minutes pass between 5:10 and 5:15?
How do you know?

 **Math at Home:** Your child practiced telling time to five minute intervals.
Activity: Ask your child to tell you when it is 6:50. Repeat with different
times to five minutes.

two hundred forty-five   **245**

3.

4.

5.

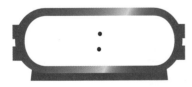   

 Spiral Review and Test Prep

Choose the correct answer.

6. How many 5-minute intervals are there in one hour?

○ 1   ○ 5   ○ 12   ○ 60

Name _____

Learn

Every 15 minutes the hand puppets sing a song. Write each time.

Math Words

quarter hour

A quarter hour is fifteen minutes.

7:00 **7:15** **7:30** **7:45**

Try it   Write each time.

1.

3:00 | : | : | :

2.

: | : | : | :

 **Sum it Up** How many minutes pass between 4:00 and 4:15? How do you know?

 Math at Home: Your child practiced telling time to the quarter hour.
Activity: Set a clock to show 5:00. Ask your child what time it will be 15 minutes later.

two hundred forty-seven **247**

Practice   Write each time.

3.

 `4:45`    `:`    `:`    `:`

4.

 `:`    `:`    `:`    `:`

Draw the minute hand to show each time.

5.    `7:00`    `7:15`   `7:30`

Mental Math

6. About how long does it take to tie your shoes?

Circle your answer.

1 minute   20 minutes

Learn

You can tell time in different ways.

It's 10:15.
It's quarter after 10.

It's 40 minutes after 8.
It's 20 minutes before 9.

Try it   Write each time more than one way.

1.

 $\underline{6} : \underline{40}$

 minutes after $\underline{6}$

$\underline{20}$ minutes before $\underline{7}$

   _____ : _____

_____ minutes after _____

_____ minutes before _____

2.

   _____ : _____

quarter after _____

   _____ : _____

quarter after _____

Sum it Up   What is a different way to say 12:15?

McGraw-Hill School Division

 Math at Home: Your child practiced telling time in different ways.
Activity: Ask your child to tell the current time in different ways. Repeat throughout the day.

Write each time.

3.

$\underline{2}:\underline{20}$

$\underline{20}$ minutes after $\underline{2}$

_____ : _____

_____ minutes after _____

4.

_____ : _____

quarter after _____ _____

_____ minutes before _____

_____ : _____

minutes after _____

_____ minutes before _____

 Problem Solving

Draw hour and minute hands to show each time.

5.

$4:47$

$8:12$

$11:22$

$1:01$

Name _____

Sequence of Events

Reading Skill   You can use sequence of events to help you solve problems.

At 2:15 the zebras dance.
At 2:45 the elephants
   prance.
At quarter after 3,
The giraffes serve tea
To you and me.

Solve.

1. What time does the show begin? __2:00__

2. Which animals perform first? _____

3. What time do the giraffes serve tea?
Write the time two ways. _____

4. Do the elephants prance before
or after the giraffes serve tea? _____

   How do you know? _____

5. At what time does the last thing happen? _____

 Math at Home: Your child used sequence of events to answer questions about a story.
Activity: Have your child show you on a clock the sequence of events in this story.

two hundred fifty-one **251**

Solve.

At 7 o'clock the hippos hop.
At quarter after 7 they
must stop.
At 20 minutes before 8,
they dance in a row.
At 8 o'clock they end
the show.

6. At what time do the hippos hop? _____

7. At what time do the hippos stop? _____

8. What do the hippos do first? _____

9. When do the hippos dance in a row?
Write the time in two ways. _____

10. How long does the hippo show last? _____

How do you know? _____

 11. Use the story.
Write a problem.

_____

_____

Check Your Progress A

Name _____

Write each time.

1.

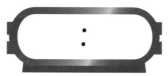

2.

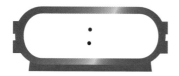

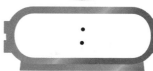

Write each time more than one way.

3.

_____ : _____

_____ minutes after _____

_____ minutes before _____

_____ : _____

_____ minutes after _____

_____ minutes before _____

Add or subtract.

1.

17	82	88	71	80	13
+ 68	+ 9	− 14	− 7	− 47	+ 66

Write each number in words.

2. 7 _____   78 _____

27 _____   70 _____

Count. Write how much money.

3. _____

TECHNOLOGY LINK

Use the Internet

- Go to www.mhschool.com/math.
- Find the site about time.
- Click on the link.
- How many years are in a decade? _____
- How many years are in 3 decades? _____

Journal Why is the Internet a good place to find information?

For more practice use Math Traveler™.

Name_____

Act It Out

Read ➤ Each puppet act takes 15 minutes. How long do 4 acts take?

What do I know?

What do I need to find out?

Plan ➤ I can use a clock to solve.

Solve ➤ 4 acts take 1 hour.

Look Back ➤ Does my answer make sense? Why?

Use a clock to act it out. Solve.

1. The puppets put on a play. Each act takes 20 minutes. How long do 3 acts take?

_____

2. The puppets dance. Each dance takes 30 minutes. How long do 2 dances take?

_____

 How do you know that 60 minutes equals 1 hour?

 Math at Home: Your child solved problems about time relationships.
Activity: Tell a simple time story. Have your child use a clock to solve.

Use a clock to act it out. Solve.

3. The puppets act out stories. Each story takes 15 minutes. How many minutes do 4 stories take to act out?

_____

4. The puppets tell jokes. Each joke is 5 minutes long. How many minutes does it take to tell 2 jokes?

_____

5. One puppet song takes 20 minutes to sing. How long do 3 puppet songs take to sing?

_____

6. One puppet story takes 15 minutes to tell. How long does it take to tell 4 puppet stories?

_____

7. The puppets dance. Each dance lasts for 15 minutes. How many minutes do 3 dances take?

_____

8. The puppets do a song and dance act. Each act is 30 minutes. The puppets do two acts. How long do 2 acts take?

_____

9. Look at the picture. Use time. Write your own problem.

_____

_____

4 ACTS
Each act is
15 minutes.
Come early!

Learn

The show starts at 1 o'clock.   It ends at 3 o'clock.

The show ends 2 hours later.

Try it   Complete the table.

Make	Start Time	End Time	Elapsed Time
1. Make puppet costumes.	2 : 00	___ : ___	___ hours
2. Make sock puppet.	___ : ___	___ : ___	___ hours

If a party starts at 3 o'clock and ends at 5 o'clock, how many hours have elapsed?

 Math at Home: Your child learned how to tell how much time has passed.
Activity: Tell your child at what time you start to prepare dinner and at what time you finish. Ask how long it took you. Repeat with other activities.

two hundred fifty-seven

McGraw-Hill School Division

Make	Start Time	End Time	Elapsed Time
3. Paint the finger puppet box.	___ : ___	___ : ___	___ hours
4. Give a puppet show.	___ : ___	___ : ___	___ hour

Draw the clock hands to show the elapsed time.

5.

 2 hours later

6.

 I hour and 30 minutes later

Learn

A calendar shows the month, weeks, days, and dates for a year.

Math Words

calendar
month
week
year

	January								February								March								April					
S	M	T	W	T	F	S		S	M	T	W	T	F	S		S	M	T	W	T	F	S		S	M	T	W	T	F	S
		1	2	3	4	5							1	2							1	2			1	2	3	4	5	6
6	7	8	9	10	11	12		3	4	5	6	7	8	9		3	4	5	6	7	8	9		7	8	9	10	11	12	13
13	14	15	16	17	18	19		10	11	12	13	14	15	16		10	11	12	13	14	15	16		14	15	16	17	18	19	20
20	21	22	23	24	25	26		17	18	19	20	21	22	23		17	18	19	20	21	22	23		21	22	23	24	25	26	27
27	28	29	30	31				24	25	26	27	28				24	25	26	27	28	29	30		28	29	30				
																31														

	May								June								July								August					
S	M	T	W	T	F	S		S	M	T	W	T	F	S		S	M	T	W	T	F	S		S	M	T	W	T	F	S
			1	2	3	4								1			1	2	3	4	5	6						1	2	3
5	6	7	8	9	10	11		2	3	4	5	6	7	8		7	8	9	10	11	12	13		4	5	6	7	8	9	10
12	13	14	15	16	17	18		9	10	11	12	13	14	15		14	15	16	17	18	19	20		11	12	13	14	15	16	17
19	20	21	22	23	24	25		16	17	18	19	20	21	22		21	22	23	24	25	26	27		18	19	20	21	22	23	24
26	27	28	29	30	31			23	24	25	26	27	28	29		28	29	30	31					25	26	27	28	29	30	31
								30																						

	September								October								November								December					
S	M	T	W	T	F	S		S	M	T	W	T	F	S		S	M	T	W	T	F	S		S	M	T	W	T	F	S
1	2	3	4	5	6	7				1	2	3	4	5							1	2		1	2	3	4	5	6	7
8	9	10	11	12	13	14		6	7	8	9	10	11	12		3	4	5	6	7	8	9		8	9	10	11	12	13	14
15	16	17	18	19	20	21		13	14	15	16	17	18	19		10	11	12	13	14	15	16		15	16	17	18	19	20	21
22	23	24	25	26	27	28		20	21	22	23	24	25	26		17	18	19	20	21	22	23		22	23	24	25	26		
29	30							27	28	29	30	31				24	25	26	27	28	29	30		29	30	31				

There are 12 months in 1 year.
There are 52 weeks in 1 year.

Try it   Use the calendar to answer each question.

1. What is the month just after August? _____

2. What is the month just before December? _____

3. How many days are in one week? _____

Sum it Up   Which months have 31 days?

_____

Math at Home: Your child practiced reading a calendar.
Activity: Point to a day on the calendar. Ask what day of the week it is. Ask how many days are in the month.

Use the calendar on page 259 to
answer each question.

4. How many days are there between March 16

and March 30? __13__ days

5. On which day of the week does February begin? _____

6. Wally says that February has four weeks.

Do you agree? Explain? _____

7. Which months are between January and April? _____

8. How many days are in each month?

January _____ July _____ October _____

9. Wally's puppet show lasted from July 14 until August 3.

How many days did the show last? _____ days

How many weeks did the show last? _____ weeks

10. The puppet theater is closed from December 21 to January 1.

How many days is the theater closed? _____ days.

11. What is the date of the third Wednesday

in October? _____

Problem Solving

12. Look at today's date.
Is today's date closer to   _____
September 1 or June 1?
Tell how you found out.   _____

Name _____

Plan a Puppet Show

How long is each act?

You want to plan a puppet show.
The show will last one hour.

Puppet Show Acts	
Elephant Ballet	20 minutes
Banjo Piggy	20 minutes
Lion Friends	10 minutes
Winter Moose	10 minutes
Rabbit Race	30 minutes
Rascal Raccoon	10 minutes
Panda in the Park	30 minutes

1. Plan your puppet show.
Choose which acts you want to have.

_____

_____

_____

_____

_____

Workspace

two hundred sixty-one **261**

2. Draw one of the acts you
   chose. How did you decide?

Workspace

_____

_____

_____

_____

3. How do you know that your
   show is one hour? Explain.

_____

_____

_____

_____

4. What if the show lasted for one and a half hours?
   What other acts could you choose?

_____

_____

_____

Name _____

How Long Is Your Shadow?

You can make a shadow of yourself outside.

You Will Need

string
scissors

What to Do

1. At 9:00 on a sunny day, find the best place to stand to make a shadow of yourself.

2. Have a partner use string to measure how long your shadow is. Cut the string. Record on the chart.

3. Stand in the same place again at 12 o'clock and 3 o'clock. Measure your shadow.

4. Compare your string measures.

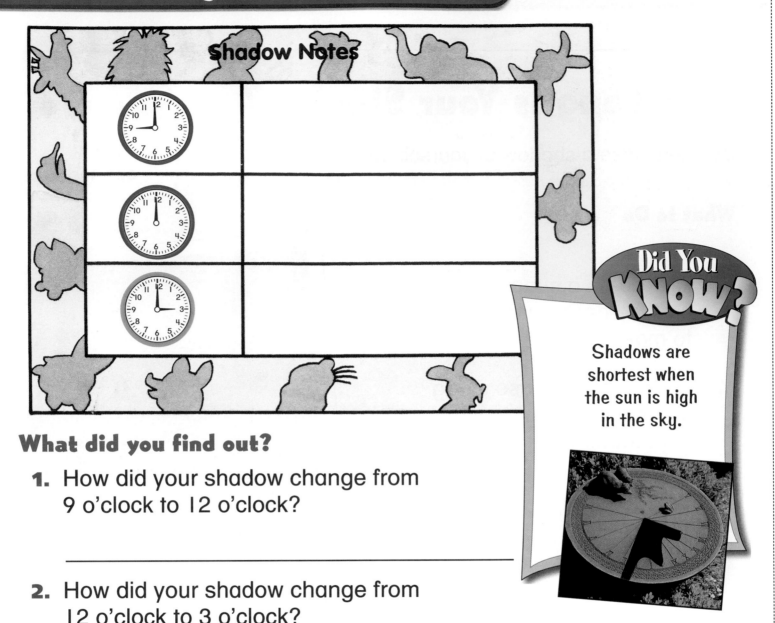

Shadow Notes

Did You KNOW?

Shadows are shortest when the sun is high in the sky.

What did you find out?

1. How did your shadow change from 9 o'clock to 12 o'clock?

   _____

2. How did your shadow change from 12 o'clock to 3 o'clock?

   _____

3. At what time was your shadow the shortest? _____

4. At what time was your shadow the longest? _____

5. At what time do you think you can make a longer shadow? Explain.

   _____

 Want to do more?

What measuring tool could you use to measure the shadow strings? Try it.

Check Your Progress B

Name _____

Use a clock to act it out. Solve.

1. Each puppet act takes 15 minutes. How long do 2 acts take?

_____

4 acts? _____

2. The puppets sing songs. Each song takes 20 minutes. How long do 2 songs take?

_____

3 songs? _____

Complete the table.

Start Time	End Time	Elapsed Time
3. ____ : ____	____ : ____	____ hours
4. ____ : ____	____ : ____	____ hours

Use the calendar to answer each question.

July						
S	M	T	W	T	F	S
	1	2	3	4	5	6
7	8	9	10	11	12	13
14	15	16	17	18	19	20
21	22	23	24	25	26	27
28	29	30	31			

5. On which day of the week does July begin?

_____

6. What is the date of the second Wednesday in July?

_____

7. How many Mondays are in July?

_____

8. What day of the week is July 4?

_____

Name _____

Curtain Time

12:15

- Cover each clock with a counter.

- You and your partner take turns.

- Remove 2 counters.

- Keep counters if the times match.

- Put counters back if they do not.

- The player with more counters wins.

7:45

3:00

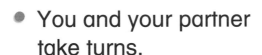

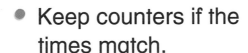

4:30

2:00

Name _____

Language and Math

Complete. Use a word from the list.

1. A _____ shows the months, weeks, days, and dates of the year.

2. One _____ is 60 seconds.

3. One _____ is 30 minutes.

Read these words.

4. One _____ is 60 minutes.

Concept and Skills

Write each time.

5.

6.

McGraw-Hill School Division

Complete the table.

7. Start Time	End Time	Elapsed Time
_____ : _____	_____ : _____	_____ hours
8. _____ : _____	_____ : _____	_____ hours

Use the calendar to complete each sentence.

9. January has _____ days.

10. January 21 is a _____.

11. The third Monday is _____.

12. The last day of January is a _____.

January						
S	M	T	W	T	F	S
		1	2	3	4	5
6	7	8	9	10	11	12
13	14	15	16	17	18	19
20	21	22	23	24	25	26
27	28	29	30	31		

Problem Solving

Use a clock to act it out. Solve.

13. Each puppet show lasts 15 minutes. How many minutes do 3 shows last?

_____ minutes

14. Each song lasts 5 minutes. How many minutes does it take to sing 3 songs?

_____ minutes

Name _____

Show each time another way.

1.

| : | 1:35 | : | 1:45 |

2.

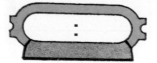    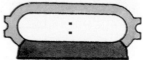

| : | 9:30 | : |

A schedule shows at what time activities happen.

After-School Schedule		
Activity	Start Time	End Time
snack	3:10	3:30
piano	3:30	4:30
homework	4:30	6:15
dinner	6:15	6:45

Mrs. Burke's class is planning a field day. Make a schedule for the field day. Write the time each activity will start and end.

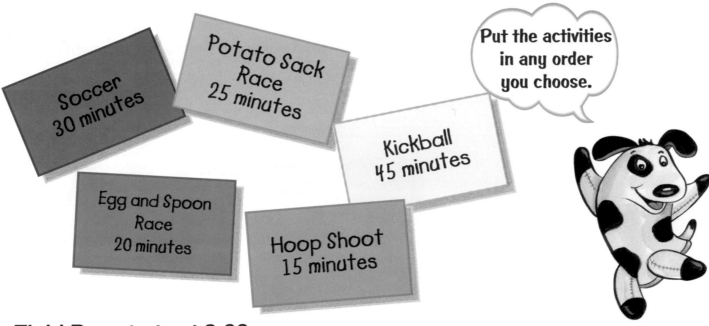

Soccer 30 minutes

Potato Sack Race 25 minutes

Put the activities in any order you choose.

Kickball 45 minutes

Egg and Spoon Race 20 minutes

Hoop Shoot 15 minutes

Field Day starts at 9:00.

Field Day Schedule		
Activity	Start Time	End Time
Potato Sack Race	9:00	

Name _____

Write each time.

1.

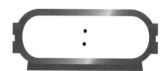

Complete.

2.

Start Time	End Time	Elapsed Time
5:00	8:00	_____

Use the calendar to answer each question.

May						
S	M	T	W	T	F	S
			1	2	3	4
5	6	7	8	9	10	11
12	13	14	15	16	17	18
19	20	21	22	23	24	25
26	27	28	29	30	31	

3. How many days are between
May 11 and May 24? _____

4. On which day of the week is May 1? _____

5. What is the date of the third Wednesday in May? _____

Solve.

6. Our class put on a puppet
show. Each show lasts 30
minutes. How long do 2
shows last?

_____

7. Each puppet tells a joke.
Each joke is 5 minutes long.
How long does it take to tell
4 jokes?

_____

Plan a puppet show that has 3 acts.
Choose a date. Mark the date on the calendar.
Choose a time. Show the time on the clock.

Then complete the poster.

You Will Need
clock
calendar

PUPPET SHOW!

January

S	M	T	W	T	F	S
			1	2	3	4	5
6	7	8	9	10	11	12
13	14	15	16	17	18	19
20	21	22	23	24	25	26
27	28	29	30	31		

Workspace

Date _____

Start Time _____

First Act
Starts _____

Second Act
Starts _____

Third Act
Starts _____

End Time _____

Show Lasts _____

Name _____

Choose the correct answer.

Mathematical Reasoning

1. What is the greatest number you can subtract from 10?

- ⬭ 0
- ⬭ 1
- ⬭ 10
- ⬭ 100

2. Diana had $1.32. She lost one coin. Now she has $1.27. Which coin did she loose?

- ⬭ penny
- ⬭ nickel
- ⬭ dime
- ⬭ quarter

Tell how you know.

_____

3. Both hands of the clock point to the 12. Which is NOT a time shown by the clock?

- ⬭ noon
- ⬭ midnight
- ⬭ 12:00
- ⬭ 12:12

Measurement and Geometry

4. Which shows a rectangle and circle?

- ⬭
- ⬭
- ⬭
- ⬭

5. Lucy plays soccer on the third day of the week. On which day does Lucy play soccer?

- ⬭ Sunday
- ⬭ Tuesday
- ⬭ Thursday
- ⬭ Friday

6. Jake is baking cookies. He needs to measure flour. Which tool should Jake use?

- ⬭ ruler
- ⬭ measuring cup
- ⬭ scale
- ⬭ clock

Algebra and Functions

7. Choose the number that makes the number sentence true.

$$35 = \boxed{} + 11$$

- ◯ 0
- ◯ 11
- ◯ 24
- ◯ 46

8. This table gives the date for three Mondays in July.

1st Monday	July 7th
2nd Monday	July 14th
3rd Monday	July 21th
4th Monday	

What date is the fourth Monday of July?

- ◯ July 22nd
- ◯ July 25th
- ◯ July 28th
- ◯ July 31st

9. Which addition sentence is related to $74 - 19 = 55$?

- ◯ $74 + 19 = 93$
- ◯ $74 + 55 = 129$
- ◯ $55 + 19 = 74$
- ◯ $74 + 19 + 55 = 148$

Number Sense

10.
$$\begin{array}{r} 73 \\ + 18 \\ \hline \end{array}$$

- ◯ 55
- ◯ 80
- ◯ 81
- ◯ 91

11.
$$\begin{array}{r} 85 \\ - 49 \\ \hline \end{array}$$

- ◯ 134
- ◯ 44
- ◯ 40
- ◯ 36

12. Which number has a 5 in the tens place?

- ◯ 5
- ◯ 25
- ◯ 51
- ◯ 23

FRANNY'S FRUIT DRINKS

TODAY'S SPECIALS
🍓 STRAWBERRY
🍊 ORANGE
🍌 BANANA

theme
All About Us

Use the Data
What ways can you show the information in the picture?

What You Will Learn
In this chapter you will learn how to:

- Read and interpret graphs.

- Record data in organized ways.

- Show the same data in different ways.

- Use a table to solve problems.

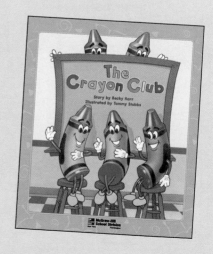

The Crayon Club
Story by Becky Harr
Illustrated by Tommy Stubbs

MATH AT HOME

Dear Family,

In Chapter 8, I will read and interpret graphs. Here are new vocabulary words and an activity that we can do together.

Math Words

pictograph

OUR FAVORITE PETS

Fish	🐟 🐟 🐟
Turtle	🐢 🐢
Dog	🐕 🐕 🐕 🐕 🐕

Each picture stands for 1 pet.

tally mark

a mark used to record data

$/ = 1$

$\cancel{||||} = 5$

bar graph

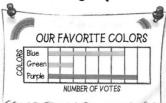

OUR FAVORITE COLORS

COLORS: Blue, Green, Purple

NUMBER OF VOTES

data

information

Guess What?

● Have your child think of a question to ask five people, such as "What is your favorite pet?"

● Then have your child record the results of the survey and discuss the information with you.

Use

pencil

paper

Additional activities at www.mhschool.com/math

McGraw-Hill School Division

Learn

pictographs
data

Math Words

Pictographs use pictures to compare information.

Cindy and her friends voted on their favorite outdoor activity.
The pictograph shows their votes.

Data is information.

Our Favorite Outdoor Activities

Each 👧 stands for 1 child.

Skating got the most votes.
How many children like skating best? __8__ children

Try it

Use the pictograph. Answer each question.

1. How many different activities does the graph show? __4__ activities

_____

2. How many children like to skate and play soccer? _____ children

_____

3. How many more children like to skate than jump rope? _____ children

_____

 How do you compare information on a pictograph?

 Math at Home: Your child read and interpreted pictographs.
Activity: Ask your child to make a pictograph of the number of plates and glasses in the cupboard.

Use the pictograph. Answer each question.

Favorite Juice

Apple		🍎
Orange		🍊
Carrot		🥕
Grape		🍇

Each 🥛 stands for 2 votes.

4. How many votes does each cup stand for?   _2 votes_

5. How many children voted?   _____ children

6. How many children voted for carrot juice?   _____ children

7. How many children voted for grape juice?   _____ children

8. How many children voted for orange juice
and grape juice?   _____ children

9. How many more children voted for carrot
juice than for grape juice?   _____ children

Problem Solving   **Algebra & functions**

Number Sense

Use data from the graph above.

10. What if each 🥛 stands for 3
votes? How many children like
carrot juice?

_____ children

11. What if each 🥛 stands for 10?
How many children like grape
juice?

_____ children

Learn

Take a survey of your class.
Ask which instrument
each student likes best.
Make a tally mark (/)
for each answer.
Complete the chart .

Math Words

survey
tally mark
chart

Favorite Musical Instruments

Instrument	Tally	Total
Drums		
Flute		
Shakers		
Trumpet		

Try it

Use data from the chart to answer
each question.

1. Which instrument does the class like best?

_____

2. Which instrument does the class like least? _____

_____

3. How many instruments does each tally mark show? _____

_____

 **Sum it Up** What do the results of your survey show?

_____

 Math at Home: Your child learned to take a survey and use tallies to record results.
Activity: Ask your child to think of a survey question to ask five people.
Then make a tally mark chart to show the responses.

two hundred seventy-nine **279**

Use data from the chart to answer each question.

> Remember 5 tally marks means 5.

4. Complete the chart. Find the total for each collection.

Favorite Collections		
Collections	Tally	Total
Stickers 🌸▽	III	3
Sports Cards 📔	ℍℍ IIII	
Toys 🎈	ℍℍ III	
Shells 🐚	IIII	

5. Which collection was chosen the most? _____

6. Which collection was chosen the least? _____

7. How many more friends liked to collect toys than shells? _____ friends

8. How many people were asked? _____ people

Use the chart.
Answer each question.

9. How many children were asked what their favorite movie is?

_____ children

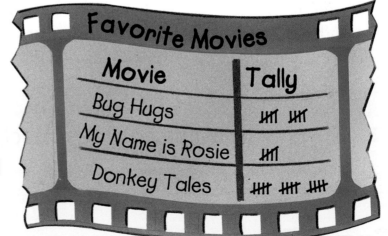

Favorite Movies	
Movie	Tally
Bug Hugs	ℍℍ ℍℍ
My Name is Rosie	ℍℍ
Donkey Tales	ℍℍ ℍℍ ℍℍ

10. Explain how you got your answer. _____

Name _____

Learn

Mr. Mack's class recorded the results of their favorite fruit pop survey on a **bar graph**.

Math Words

bar graph

Favorite Fruit Pop Flavors

Grape
Strawberry
Mango
Cherry

0 1 2 3 4 5 6 7 8 9 10
Number

What is the favorite fruit pop flavor?

strawberry

Try it

Use data from the bar graph to answer each question.

1. What is the title of the bar graph? _____

2. Does the graph show that the class liked grape or mango better? _____

3. Which flavor does the class like least? _____

4. Were the class totals for grape and cherry equal to the class total for strawberry? _____

Sum it Up

On a bar graph, what does the longest bar show?

Math at Home: Your child practiced reading, interpreting and making bar graphs.
Activity: Ask your child to explain the graph on this page to you.

two hundred eighty-one **281**

Practice

Donna surveyed her friends to find out their favorite sea animals. 10 people liked seals best. 3 people liked sea lions. 5 people liked whales. 2 people said dolphins were their favorite.
Show the votes on the bar graph.

Color one space for each vote.

Our Favorite Sea Animals

Kind of Animal		0	1	2	3	4	5	6	7	8	9	10
	seals											
	sea lions											
	whales											
	dolphins											

Number of Votes

Use data from the graph to answer each question.

5. Which animal is the class favorite? ___seal___

6. How many more children like seals than whales? _____ children

7. How many children voted in all? _____ children

Spiral Review and Test Prep

Test Prep

Choose the correct answer.

8. How many children like cereal?

- ◯ 10
- ◯ 9
- ◯ 8
- ◯ 7

Favorite Breakfast

9. Choose the number that makes the number sentence true.

$$56 = \boxed{} + 10$$

- ◯ 46
- ◯ 36
- ◯ 26
- ◯ 16

Name _____

Draw Conclusions

Reading Skill   You can use diagrams to show data and draw conclusions.

The second graders collected this data about the third-grade class.

Do you like video games or board games?

Maria
Andy
Chris
Tommy

Juan
Leslie
Toby
Lisa

Betty
Pam
Pete
Brent
Susie

Like Video Games   Like Both   Like Board Games

Answer each question.

1. How many children like board games?   ___9___ children

2. How many children like video games?   _____ children

3. How many children were asked which games they like?   _____children

4. How many children like board games but not video games?   _____ children

 Math at Home: Your child used diagrams to solve problems and draw conclusions.
Activity: Ask your child to explain the diagram on this page.

Solve.

The second graders collected this data about the first-grade class.

Do you like soccer or kickball?

Do you like soccer or kickball?

Like Soccer	Like Both	Like Kickball
Carmen	Bill	Barbra
Alan	Sheri	Stu
	Terry	Juan
Bobby	Loren	Harry
	Stan	
Lisa	Beth	Alice

Answer each question.

5. How many children like soccer and kickball?   _6_ children

6. How many children like soccer?   _____ children

7. How many children were asked which games they like?   _____ children

8. How many children like kickball but not soccer?   _____ children

9. Would more children rather play kickball or soccer? Explain your thinking.

_____

Name _____

Use data from the pictograph to answer each question.

FAVORITE COOKIES

Oatmeal	
Vanilla	
Peanut Butter	
Raisin	

Each 🍪 stands for 1 child.

1. What do the numbers on the graph show?

_____

2. How many children does each cookie stand for?

_____ child

3. How many children voted for raisin cookies?

_____ children

4. Was the number of votes for raisin cookies equal to the number of votes for all

the other cookies? _____

5. How many children voted for oatmeal cookies?

_____ children

6. How many more children voted for raisin than

oatmeal? _____ children

7. How many children voted

in all? _____ children

Use the graph to answer each question.

8. What do the bars tell you?

_____

9. Which story got the most votes?

_____

10. What begins each row?

_____

OUR FAVORITE STORIES

KIND OF STORY

Sports
Animals
Folk Tales
Real Stories

0 1 2 3 4 5
NUMBER OF VOTES

Add or subtract.

1.
$$\begin{array}{r} 82 \\ +\ 9 \\ \hline \end{array} \qquad \begin{array}{r} 26 \\ -18 \\ \hline \end{array} \qquad \begin{array}{r} 44 \\ +28 \\ \hline \end{array} \qquad \begin{array}{r} 75 \\ -46 \\ \hline \end{array} \qquad \begin{array}{r} 90 \\ -\ 5 \\ \hline \end{array} \qquad \begin{array}{r} 47 \\ +17 \\ \hline \end{array}$$

Write each time.

2.

_____ : _____   _____ : _____   _____ : _____   _____ : _____

TECHNOLOGY LINK

Use a Table to Make a Graph

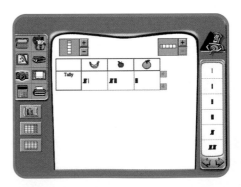

- Use a table.
- Choose a mat to show tallies.
- Add 2 columns. Label the columns.
- Stamp the tally marks.
- Click the graph key. Click on Tally.
- Which has the longest bar? _____

Make a table. Show tally marks for 3 rabbits, 5 cats, and 8 dogs. Make a graph. Which has the longest bar? _____

For more practice use Math Traveler™.

Name _____

Make a Table

Read ▶ Eliza saves money every week to buy books. In the first week of April, she saved $3. Next week she saved $2. In the third week of April she saved $4, and in the fourth week she saved $3. During which week did Eliza save the most money?

Eliza's Book Money

Week	1	2	3	4
How Much	$3	$2	$4	$3

Plan ▶ I can make a table to solve.

Solve ▶ Eliza saved the most money during the third week.

Look Back ▶ Does my answer make sense? Why?

Fill in the table. Then answer each question.

Tim wants to buy a new puzzle. He needs to save $10. During the first week, he saved $2. In the second week he saved $1. He saved $5 in the third week. In the fourth week he saved $4.

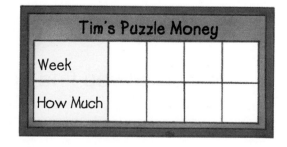

Tim's Puzzle Money

Week				
How Much				

1. How much money did Tim save in all? _____

2. During which week did Tim save the most money? _____

 How can a table help you solve problems?

 Math at Home: Your child used a table to solve problems.
Activity: Ask your child to make a table to show how many eggs there are in 3 egg cartons.

Fill in the table. Then answer the questions.

Mrs. Marshall's second grade class is holding a book drive. They collected 10 books in January and 12 in February. In March they collected another 10. In April they collected 22, and in May they collected 20.

Second Grade Book Collection

Month	Jan.	Feb.	March	April	May
How Many					

1. During which month were the most books collected? _____

2. During which two months were the same number of books collected? _____

3. How many more books were collected in May than in February? _____ books

4. How could you find how many books were collected in all?

_____

Critical Thinking Journal

5. Look at the table. Write a problem.

_____

Ben's Baseball Card Money

Week	1	2	3	4
How Much	$0	$4	$3	$5

Learn

You can show data in more than one way.

Games

Relay race	IIII IIII
Giant steps	IIII III
Kickball	IIII II
Tag	IIII

Table

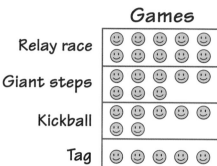

Pictograph

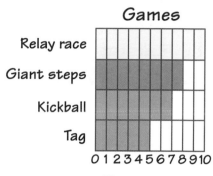

Bar graph

Jill can show the same data in a table, pictograph, or bar graph.

Try it   Answer each question.

1. How is a table like a pictograph and bar graph? How is it different?

_____

_____

2. Which game got 8 votes? _____

3. From which data display did you get your answer for question 2? Explain why.

_____

Sum it Up   How can you show the same data in different ways?

McGraw-Hill School Division

Math at Home: Your child showed data in different ways.
Activity: Ask your child to record how many forks, spoons, and knives there are on the dinner table. Then ask him or her to show the data in two different ways.

Use data from the tally table to make the bar graph. Answer each question.

4.

Picnic Snacks	
Corn Chips	I
Popcorn	|||| |||
Raisins	|||| |||| ||

Picnic Snacks

Snacks: Corn Chips, Popcorn, Raisins

Number of Votes: 0 1 2 3 4 5 6 7 8 9 10 11 12

5. Which snack got 8 votes? _popcorn_

6. From which data display did you get your answer for question 5? Explain.

_____

7. Which snack got the most votes? _____

8. From which data display did you get your answer for question 7? Explain.

_____

Critical Thinking Journal

9. If you go shopping for the picnic snacks, which data display do you think is easier to use? Explain.

_____

10. If you compare data about picnic snacks, which data display do you think is easier to use? Explain.

_____

Learn

Tracy asked 5 friends to tell their
favorite numbers.
Here is the set of data.

Math Words
set
range
mode

2   7   10   36   7

The range is the difference
between the greatest and
least numbers.
Subtract to find the range.

$36 - 2 = 34$

The range is _34_.

The mode is the number that
occurs most often.

7 occurs most often.

The mode is _7_.

Try it   Answer each question.

Tom scored these goals in 5 soccer games.

1   8   4   4   6

1. What are the greatest and least numbers? _8, 1_

2. What is the range of the numbers? _____

3. Which number is shown more than once? _____

4. What is the mode of the numbers? _____

Sum it Up How do you find the range for a set of data?

Math at Home: Your child practiced finding the range and mode of a set of data.
Activity: Ask your child to record the age for each family member. Ask him
or her to tell you what the range is. Is there a mode? What is it?

two hundred ninety-one **291**

McGraw-Hill School Division

Use the data. Answer each question.

The post office sold this many postcard stamps in the last five days.

15 28 15 14 34

5. What is the range of the numbers? __20__

6. What is the mode of the numbers? _____

The nature museum sold this many tickets in the last 4 days.

17 42 22 22

7. What is the range of the numbers? _____

8. What is the mode of the numbers? _____

Spiral Review and Test Prep

Test Prep

Choose the correct answer.

9. What is the mode for this set of data?

19, 20, 5, 18, 19

- ◯ 3
- ◯ 5
- ◯ 19
- ◯ 25

10. Jan has 49¢. Which group of coins does she have?

- ◯ 4 dimes, 4 pennies
- ◯ 8 nickels, 4 pennies
- ◯ 1 quarter, 2 dimes, 4 pennies
- ◯ 1 quarter, 2 dimes, 9 pennies

Name _____

Make a Picture

You are painting a mural that shows favorite school activities. Use the data to help you.

Favorite Activities	
Music	ЖН I
Art	ЖН ЖН II
Reading	ЖН II
Soccer	IIII

You Decide!

1. Decide what activities you want to paint on the mural.

2. Make a list.

_____

_____

_____

_____

_____

_____

_____

Workspace

Problem Solving • Decision Making

3. Tell about how you made your decision.

_____

_____

4. Draw a picture of the activities you want
to show on the mural.

Workspace

5. What if you decide to use
the information on the
mural to make a graph?
What kind of graph could
you make?

Workspace

_____

Draw the graph.

Name _____

What Is Your Favorite Season?

You can make a graph to show the favorite season of your class.

Fall   Winter   Spring   Summer

What to do

1. Survey the class.

2. Decide on a symbol to use for each season.

3. Record the data.

4. Decide how you want to display the data.

5. Make the display.

Our Favorite Season	
🍂	
❄️	
🌳	
🌼	

What did you find out?

1. What can you tell from looking at the information?

_____

2. Does your class like spring or fall better? How do you know?

_____

3. What season does your class like best?

_____

4. How else could you show the information? Explain.

_____

Want to do more?

How does the weather change from season to season?
How can you find out?

Name _____

Use data from the tally table to make the bar graph.
Then answer each question.

1.

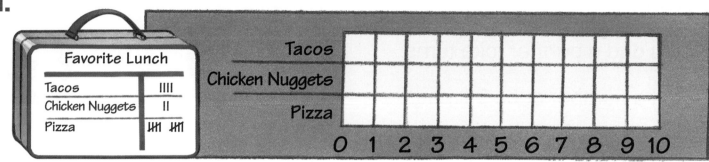

Favorite Lunch	
Tacos	IIII
Chicken Nuggets	II
Pizza	̶H̶H̶ ̶H̶H̶

2. Which lunch got 4 votes? _____

3. Which lunch got the least votes? _____

4. Which lunch got the most votes? _____

5. How many children voted in all? _____ children

6. How many more children voted
for pizza than voted for tacos? _____ children

The dance sold this many tickets in the last five days.

Dance Tickets Sold	
Monday	25
Tuesday	18
Wednesday	25
Thursday	19
Friday	14

7. What is the range of the numbers? _____

8. What is the mode of the numbers? _____

Use data from the table to answer each question.

9. Does Denise have enough money
to buy a game that costs $12? _____

10. How much money did
Denise save in all? _____

Denise's Game Money				
Week	1	2	3	4
How Much	$3	$4	$3	$3

Name _____

Tally Ho!

You Will Need

3

- You and a partner take turns.

- Roll a number cube 5 times.

- Make a tally next to the number you roll.

- Now find the range.

- The player with the highest number wins.

- Play again.

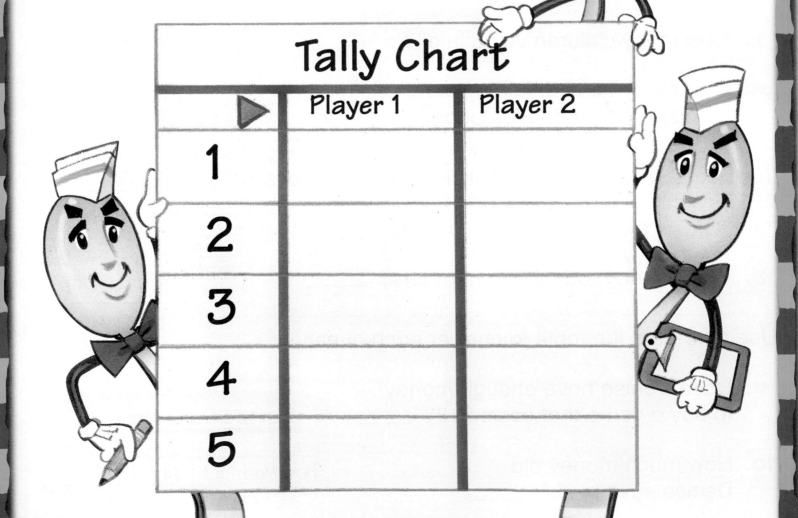

Tally Chart

▶	Player 1	Player 2
1		
2		
3		
4		
5		

Name _____

Language and Math

Complete. Use a word from the list.

1. A _____ uses pictures or drawings to show how many.

2. A _____ uses bars to show how many.

3. The _____ is the number that occurs most often in a set of data.

4. The _____ is difference between the greatest and least number in a set of data.

Concepts and Skills

Use data from the graph to answer each question.

Our Favorite Drinks

Kind of Drink	0	1	2	3	4	5	6	7	8	9	10
Milk											
Orange Juice											
Apple Juice											
Grape											

Number of Votes

5. Which drink got 8 votes? _____

6. Which drink got 2 more votes than grape? _____

7. How many children voted altogether? _____ children

Use data from the pictograph to answer each question.

Our Favorite Shows

Adventure	▢ ▢ ▢
Funny shows	▢ ▢ ▢ ▢ ▢ ▢ ▢
Animal	▢ ▢ ▢ ▢ ▢
Game	▢ ▢

8. How many votes did animal shows get? _____ votes

9. What is the favorite show?

_____

10. How many people voted altogether?

_____ people

Use data from the tally chart. Then answer the questions.

Our Favorite Summer Activity

Swimming	ЖЖ |||
Camping	ЖЖ ||
Vacation	||||
Hiking	|||

11. Did hiking get more votes than camping?

_____

12. How many people voted for swimming? _____ people

13. What is the activity that got the fewest votes? _____

Problem Solving

Use data from the table to answer each question.

Lucy's Book Money

Week	1	2	3	4
How much?	$2	$1	$3	$3

14. Does Lucy have enough money to buy a book that costs $10? _____

15. How much money did Lucy save in all? _____

Name _____

Mrs. Kelly's class voted for their favorite picnic food. These are the results.

Picnic Food		
	Burgers	12 votes
	Chicken Wings	10 votes
	Fruit Salad	6 votes
	Corn On The Cob	8 votes

Use the data from the poster to complete the bar graph.

1. What is the title of the graph? _____

2. Which food got the most votes? _____

3. How many children voted for corn on the cob

   and fruit salad? _____ children

4. How many more children voted for chicken

   wings than for corn on the cob? _____ children

5. How many children voted in all? _____

6. What is a question this graph can answer? _____

Mr. Cole's class went to the amusement park.
The children rode on these rides.
They made this graph.

AMUSEMENT PARK RIDES

	Roller Coaster	Ferris Wheel	Bumper Cars	Fun House

1. How many children rode on the roller coaster? _____ children

2. How many more children rode on the Ferris wheel than the bumper cars? _____ children

3. There are 25 children in the class. Did some children go on more than one ride?
How do you know?

_____

4. What is another question this graph can answer?

_____

Name _____

Use data from the tally chart to make a bar graph.
Then answer the questions.

1.

2. What is the title of the bar graph? _____

3. Which story got the most votes? _____

4. How many more votes did *Rabbit and Me*

get than *Turtle's Race*? _____

5. Did any story get more than 12 votes? _____

6. How many children voted altogether? _____ children

Use these numbers to answer each question: 18, 20, 20, 25, 26, 28.

7. What is the mode of these numbers? _____

8. What is the range of the numbers? _____

Use data from the table to answer each question.

9. Can Steven buy a $10 book? _____

10. How much did Steven save in all? _____

Steven's Book Money				
Week	1	2	3	4
How Much	$2	$1	$4	$2

SDAP1.2, 1.3, 1.4

Spin the spinner 12 times.
Record the results of each spin in the tally chart.

Color	Number of Spins
Red	
Blue	
Green	
Yellow	

Use data from your tally chart to make a bar graph.
Then answer the questions.

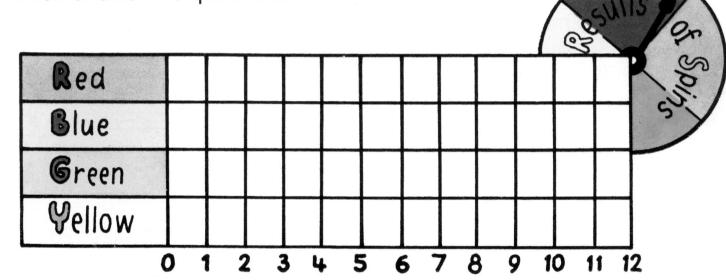

Red
Blue
Green
Yellow

0 1 2 3 4 5 6 7 8 9 10 11 12

1. Which color did you get the most? _____

2. How many times did you spin red? _____

3. What is another question this
graph can answer? _____

Portfolio

You may want to put this page in your portfolio.

Name _____

Choose the correct answer.

Statistics, Data Analysis, and Probability

1. What is the mode of the numbers?

   13, 23, 2, 5, 5, 1
   - ◯ 1
   - ◯ 5
   - ◯ 13
   - ◯ 23

   Use the graph to answer questions 2 and 3.

FAVORITE COLOR	
RED	☺ ☺ ☺
BLUE	☺ ☺
GREEN	☺
YELLOW	☺ ☺ ☺ ☺

   EACH ☺ STANDS FOR 2 VOTES.

2. Which color got 2 votes?
   - ◯ Red
   - ◯ Green
   - ◯ Blue
   - ◯ Yellow

3. How many more votes did yellow get than blue?
   - ◯ 1
   - ◯ 2
   - ◯ 3
   - ◯ 4

Number Sense

4.
$$\begin{array}{r} 38 \\ + 49 \\ \hline \end{array}$$
   - ◯ 47
   - ◯ 67
   - ◯ 77
   - ◯ 87

5. There are 72 oranges in the box. The clerk takes out 27 oranges to make juice. How many oranges are left?
   - ◯ 45
   - ◯ 55
   - ◯ 55
   - ◯ 99

   How did you get your answer?

   _____

6. Which are odd numbers?
   - ◯ 83, 84, 85, 86
   - ◯ 76, 79, 82, 85
   - ◯ 51, 55, 57, 59
   - ◯ 66, 68, 70, 71

Algebra and Functions

7. Which number completes the addition?

$$\boxed{} + 34 = 62$$

- ⬭ 96
- ⬭ 62
- ⬭ 32
- ⬭ 28

8. Which is the turnaround fact for $43 + 19 = 62$?

- ⬭ $19 + 43 = 62$
- ⬭ $62 + 19 = 81$
- ⬭ $43 - 19 = 24$
- ⬭ $62 - 19 = 43$

9. Which numbers make this sentence true?

$$2 \text{ hours} = \boxed{} + \boxed{} \text{ minutes}$$

- ⬭ $1 + 1$
- ⬭ $10 + 10$
- ⬭ $30 + 30$
- ⬭ $60 + 60$

Mathematical Reasoning

10. Henry has a quarter and 3 pennies. He found 6 more pennies. How much money does Henry have now?

- ⬭ 25 ¢
- ⬭ 27 ¢
- ⬭ 28 ¢
- ⬭ 34 ¢

11. Sue started her homework at 3:30. She finished at 5:30. How long did she spend doing homework?

- ⬭ 1 hour
- ⬭ 2 hours
- ⬭ $1\frac{1}{2}$ hours
- ⬭ 3 hours

Explain your reasoning.

---

12. Kim bought a large bag of 36 apples. She left 28 apples at home and took the rest to school to share. How many apples did she share at school?

_____ apples

Measurement

Long, Long Ago

Use the Data

What are some other ways that you could measure the dinosaur foot?

What You Will Learn

In this chapter you will learn how to:

- Measure length, weight, and capacity.

- Find the perimeter and area of a figure.

- Read temperatures.

- Guess and test to solve problems.

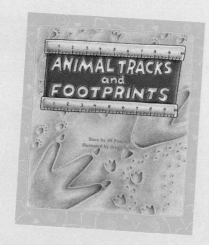

ANIMAL TRACKS and FOOTPRINTS

Story by Jill Penteford
Illustrated by Oscar Volkery

McGraw-Hill School Division

Math Words

area

The area is 6 square units.

perimeter

2"

1"

The perimeter is 6 inches.

temperature

The temperature is 80°.

length

how long something is

Dear Family,

In Chapter 9, I will learn about measuring in different ways. Here are new vocabulary words and an activity that we can do together.

A Measurement Game

● Have your child measure the length of the spoons to the nearest inch.

● After your child measures the spoons, tell your child the length of one of the forks, and have your child estimate which fork is that length. Then your child can measure and check.

use

different length spoons and forks

inch ruler

Additional activities at www.mhschool.com/math

Learn

You can use a ruler to measure.
Measure the length of
the dinosaur.

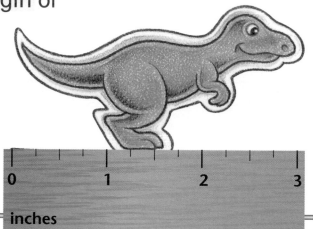

Math Words

foot [ft]
inch [in]
length
measure

12 inches
equals 1 foot.

The dinosaur
is about 3
inches long.

inches

Try it

Estimate. Then use an inch ruler to measure.

	Estimate	Measure
1.	about __3__ inches	__3__ inches
2.	about _____ inches	_____ inches
3.	about _____ inches	_____ inches
4.	about _____ inches	_____ inches

Sum it Up

How do you use an inch ruler to measure?

Math at Home: Your child used a ruler to measure inches.
Activity: Have your child show you how to measure the length of a crayon using
an inch ruler.

McGraw-Hill School Division

Find objects like the ones shown.
Estimate the length or height.
Use a ruler to measure.

12 inches = 1 foot

Object	Estimate	Measure
5. ERASER	about _____ inches	_____ inches
6.	about _____ inches	_____ inches
7.	about _____ inches	_____ inches
8.	about _____ inches	_____ inches

Mental Math

9. Timmy is 4 feet tall.
   His brother is 2 feet taller.
   How tall is Timmy's brother? _____ feet

10. Laura is 5 feet tall.
    Her sister is 2 feet shorter.
    How tall is Laura's sister? _____ feet

Learn

You can use a yardstick to measure the length of the dinosaur.

The dinosaur is about 24 inches long.

24 inches is the same as 2 feet.

Math Word

yard [yd]

12 inches equals 1 foot. 3 feet equals 1 yard.

Try it

Find objects like the ones shown. Estimate the length or height. Measure and record.

Object	Estimate	Measure
1.	about _____	_____
2.	about _____	_____
3.	about _____	_____
4.	about _____	_____

Sum it Up

How many inches are in 3 feet? How do you know?

Math at Home: Your child learned about inches, feet, and yards.
Activity: Have your child show you some items in your home that are about 1 inch, 1 foot, and 1 yard long.

Estimate.
Find an object about this long.

12 inches equals
1 foot.
3 feet equals 1 yard.

Estimate.	Name your object.	Measure.
5. about 1 inch	_____	_____
6. about 8 inches	_____	_____
7. about 1 foot	_____	_____
8. about 2 feet	_____	_____
9. about 1 yard	_____	_____

Spiral Review and Test Prep

Choose the correct answer. Fill in the ⬭.

10. The length of a real
pencil is about _____.

- ⬭ 1 inch
- ⬭ 5 inches
- ⬭ 5 feet
- ⬭ 5 yards

11. What is the mode for this
set of data?
15 8 7 15 9

- ⬭ 7
- ⬭ 8
- ⬭ 9
- ⬭ 15

Learn

Math Words

cup (c)
pint (pt)
quart (qt)

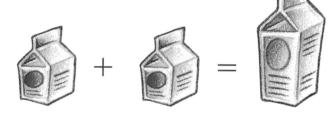

cup + cup = pint

pint + pint = quart

Try it   Complete.

1. __2__ cups = 1 pint

2. _____ pints = 1 quart

3. _____ cups = 1 quart

4. _____ cups = 2 quarts

5. _____ pints = 2 quarts

6. _____ quarts = 4 cups

Sum it Up How many cups are in 2 pints? How do you know?

Math at Home: Your child learned about cups, pints, and quarts.
Activity: Have your child show you some containers
in your home that hold about 1 cup, 1 pint, and 1 quart.

Practice   Circle the better estimate.

7.

more than I cup

(less than I cup)

8.

more than I cup

less than I cup

9.

more than I pint

less than I pint

10.

more than I pint

less than I pint

11.

more than I quart

less than I quart

12.

more than I quart

less than I quart

13.

more than I quart

less than I quart

14.

more than I quart

less than I quart

Algebra & functions   Complete the table.

quarts		I	2	3	4
pints		2			
cups		4			

Name _____

Learn

An ounce and a pound are units of measure of weight.

Math Words

ounce (oz)
pound (lb)

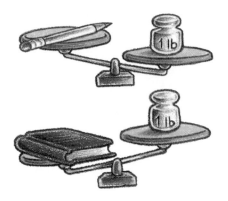

The pencil weighs less than 1 pound.
The book weighs more than 1 pound.

The paper clip weighs less than 1 ounce.
The toy weighs more than 1 ounce.

16 ounces equals 1 pound.

Try it

Circle the better estimate. Then use a balance and weights to check.

1.

more than 1 pound

less than 1 pound

more than 1 ounce

less than 1 ounce

2.

more than 1 pound

less than 1 pound

more than 1 ounce

less than 1 ounce

 Sum it Up

How do you know when an object weighs more than 1 pound?

 **Math at Home:** Your child determined whether objects weigh more or less than a pound.
Activity: Have your child show you an object in your home that weighs more than a pound and an object that weighs less than a pound.

Practice — Circle the better estimate.

3.

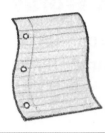
more than I ounce

(less than I ounce)

more than I pound

less than I pound

4.

more than I ounce

less than I ounce

more than I pound

less than I pound

5.

more than I pound

less than I pound

more than I ounce

less than I ounce

Algebra & functions

6. How many eggs will balance the scale?

4 ◯ = 8 ◯   2 ◯ = ___ ◯

Learn

The distance around a figure is the perimeter. What is the perimeter of this shape?

Math Words

perimeter

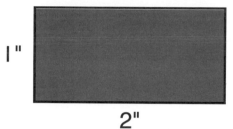

We are measuring the perimeter.

2"

1"   1"

2"

<u>2</u> + <u>1</u> + <u>2</u> + <u>1</u> = <u>6</u> inches

The perimeter is <u>6</u> inches.

Try it

Find objects like the ones shown. Use an inch ruler. Find the perimeter.

1. ____ + ____ + ____ + ____ = ____ inches

2. ____ + ____ + ____ + ____ = ____ inches

3. ____ + ____ + ____ + ____ = ____ inches

Sum it Up

How do you find the perimeter of a sheet of paper?

 **Math at Home:** Your child found the perimeter of objects.
Activity: Have your child find the perimeter of a book in your home.

McGraw-Hill School Division

Find the perimeter of each figure.
Use an inch ruler.

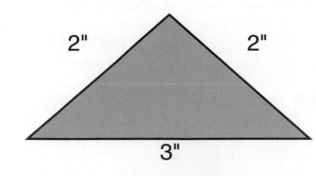

The distance around
a figure is called
the perimeter.

4.

2"   2"

3"

__2__ + __2__ + __3__ = __7__ inches

5.

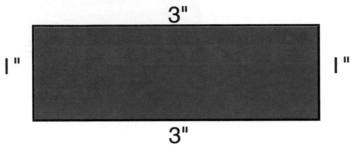

3"

1"   1"

3"

_____ + _____ + _____ + _____ = _____ inches

6.

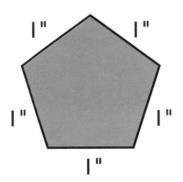

1"   1"

1"   1"

1"

_____ + _____ + _____ + _____ + _____ = _____ inches

Problem Solving

7. Draw 2 different figures
that have the same
perimeter.
Use a ruler.

Learn

Math Words

area

Area is measured in square units.

_____6_____ square units

Try it

Find the area of each figure.
Color squares or use paper squares to cover each figure.

1.

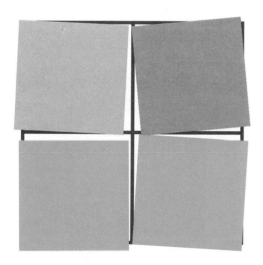

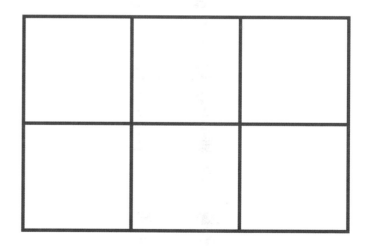

_____4_____ square units   _____ square units

Sum it Up

How do you find the area of an object?

Math at Home: Your child learned about area.
Activity: Have your child find the area of a rectangular pan using square
units, such as square crackers.

Color to show the number of square units.

2.

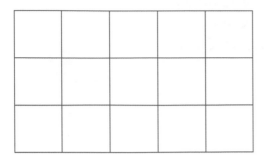

6 square units

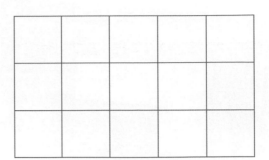

9 square units

3.

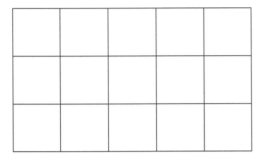

10 square units

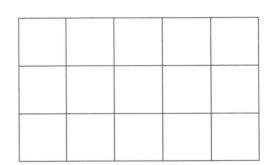

12 square units

Number Sense

4. Estimate the area of the dinosaur's footprint.

About _____ square units

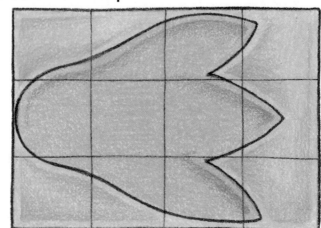

Name _____

Use Maps

Reading Skill   You can use a map to solve problems.
Read the map.

Use your inch ruler to measure. Solve.

1. How long is the path from the cave to the trees? __2 in.__

2. How long is the path from the trees to the lake? _____

3. How long is the path from the cave to the trees to the lake?

_____

4. Draw a shorter path from the cave to the lake. How long is it?

_____

Math at Home: Your child learned to solve math problems by using a map.
Activity: Show your child a map and talk about how you use it.

Use your ruler to measure. Solve.

5. How long is the path from the cave to the river? _____

6. How long is the path from the river to the rock? _____

7. How long is the red path from
the cave to the mountain? _____

8. How long is the blue path from
the cave to the mountain? _____

9. Draw a shorter path from
the cave to the mountain. How long is it? _____

Name _____

Estimate. Then measure.

	Estimate	Measure

1.  about _____ inches _____ inches

2.  about _____ inches _____ inches

Circle the better estimate.

3.

more than 1 cup

less than 1 cup

4.

more than 1 cup

less than 1 cup

5.

more than 1 ounce

less than 1 ounce

6.

more than 1 pound

less than 1 pound

Find the perimeter.

7.

_____ + _____ + _____ + _____

The perimeter is about _____ inches.

Color to show the number of square units.

8.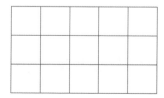

4 square units

9.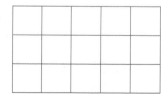

6 square units

10.

9 square units

Add or subtract.

1. | 27 | 82 | 64 | 47 | 53 | 38 |
    | + 35 | − 18 | + 32 | − 35 | − 27 | + 24 |

Draw the clock hands to show the end time.

2.    2 hours later

3.    3 hours later

TECHNOLOGY LINK

Use the Internet

- Go to www.mhschool.com/math.

- Find the site about measuring.

- Click on the link.

- How many inches are in 1 span? _____

- How many inches are in 2 spans? _____

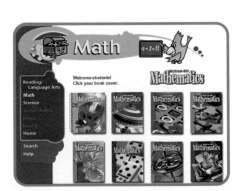

 Why is the Internet a good place to find information?

For more practice use Math Traveler™.

Name _____

Guess and Check

What do I know?

Read ➤ How long is the path?

What do I need to find out?

Plan ➤ I can estimate the length.
Then I can check my estimate with a ruler.

Solve ➤ I estimate 5 inches. When I measure,
I find the path is 6 inches long.

Look Back ➤ Does my answer make sense? Why?

How long is each path?
Estimate the length. Then use a ruler to check.

1.

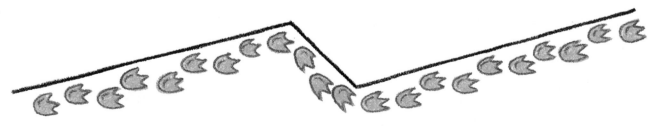

Estimate __7__ inches   Check __7__ inches

 Do your estimates make sense? Explain.

McGraw-Hill School Division

 Math at Home: Your child used the strategy Guess and Check to solve problems.
Activity: Give your child some items that are less than a foot in length. Have your child
estimate the length of each item and then check the estimate by
measuring with a ruler.

three hundred twenty-five **325**

How long is each path?
Estimate the length. Then use a ruler to check.

2.

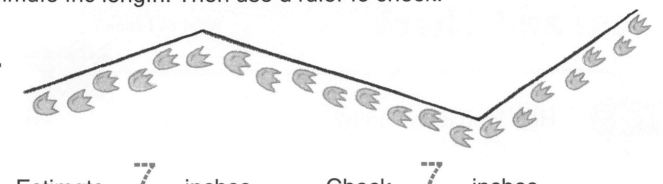

Estimate ___7___ inches   Check ___7___ inches

3.

Estimate _____ inches   Check _____ inches

4.

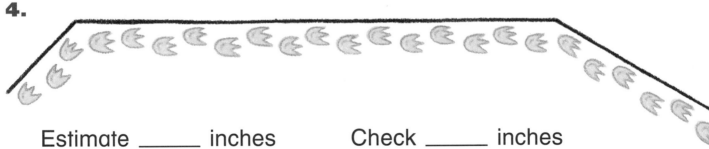

Estimate _____ inches   Check _____ inches

5.

Estimate _____ inches   Check _____ inches

Critical Thinking   Journal   **Workspace**

6. Draw a path with 3 parts.
   Estimate the length of the path.
   Then use a ruler to check.

Name _____

Learn

You can use a **centimeter ruler** to measure.

Math Words
centimeter (cm)
meter (m)

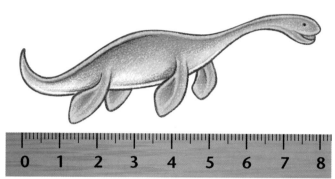

There are 100 centimeters in 1 meter.

This dinosaur is about 8 centimeters long.

Try it   Use a centimeter ruler to measure.

1.

___6___ centimeters

2.

_____ centimeters

 **Sum it Up!**   How do you use a centimeter ruler to measure the length of the dinosaur?

 Math at Home: Your child learned about centimeters and meters.
Activity: Have your child show you how to measure a toy car using a centimeter ruler.

McGraw-Hill School Division

3.

_____8_____ centimeters

4.

_____ centimeters

5.

_____ centimeters

6.

_____ centimeters

Mental Math

7. A baseball bat is about 1 meter long. Name some other items that are about 1 meter long.

_____

_____

Learn

A gram is a unit of measure of weight.
There are 1,000 grams in 1 kilogram.

Math Words

gram (g)
kilogram (kg)

less than
1 kilogram

more than
1 kilogram

about the same as
1 kilogram

Try it   Circle your estimate.

1.

lighter than 1 kilogram

(heavier than 1 kilogram)

about the same as 1 kilogram

lighter than 1 kilogram

heavier than 1 kilogram

about the same as 1 kilogram

2.

lighter than 1 kilogram

heavier than 1 kilogram

about the same as 1 kilogram

lighter than 1 kilogram

heavier than 1 kilogram

about the same as 1 kilogram

Sum it Up   Name some objects that are lighter than 1 kilogram.

Math at Home: Your child learned about grams and kilograms.
Activity: Have your child tell one object that is less than a kilogram
and one object that is more than a kilogram.

3.

lighter than I kilogram   lighter than I kilogram

(heavier than I kilogram)   heavier than I kilogram

about the same as I kilogram   about the same as I kilogram

4.

lighter than I kilogram   lighter than I kilogram

heavier than I kilogram   heavier than I kilogram

about the same as I kilogram   about the same as I kilogram

5.

lighter than I kilogram   lighter than I kilogram

heavier than I kilogram   heavier than I kilogram

about the same as I kilogram   about the same as I kilogram

Problem Solving

Mental Math

6. Which is heavier,
I kilogram of rocks or
I kilogram of feathers?
Explain.

Learn

This bottle holds 1 liter.

Try it Circle the better estimate.

1.

 more than 1 liter

(less than 1 liter)

 more than 1 liter

less than 1 liter

2.

 more than 1 liter

less than 1 liter

 more than 1 liter

less than 1 liter

Sum it Up! Does a drinking glass hold more or less than 1 liter?

 Math at Home: Your child learned about liters.
Activity: Have your child tell you one container that holds more than
1 liter and one container that holds less than 1 liter.

three hundred thirty-one **331**

3.

about 5 liters

(about 500 liters)

about 4 liters

about 400 liters

4.

about 1 liter

about 15 liters

about 1 liter

about 80 liters

5.

about 2 liters

about 15 liters

about 4 liters

about 40 liters

 Problem Solving

6. Donna filled 5 glasses with 1 liter of juice. How many glasses could she fill with

3 liters of juice? _____ liters

How did you find out? Draw a picture.

Workspace

Learn

Temperature can be measured in degrees Fahrenheit (°F) or degrees Celsius (°C).

Math Words

temperature
degrees Celsius (°C)
degrees Fahrenheit (°F)

This thermometer shows 48 °F and 3 °C.

Try it   Write each temperature.

1.

90 °F

_____ °F

_____ °C

 **Sum it Up**   How do you know the number of degrees a thermometer shows?

🏠 **Math at Home:** Your child learned about temperature.
Activity: Ask your child what you should wear when the temperature is 25°F.

2.

_|8_ °F

_____ °F

_____ °F

3.

_____ °C

_____ °C

_____ °C

Critical Thinking

4. Draw a picture. Show what you would wear to go outside.

Workspace

Learn

Darcy wants to measure the length of the dinosaur. Which tool should she use?

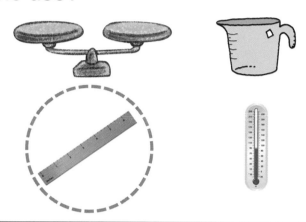

Try it
Circle the tool you would use to measure.

1. How heavy is a book?

2. How much water is in a glass?

3. How warm is it outside?

 How many different ways can you measure a pitcher?

 Math at Home: Your child learned about measurement tools.
Activity: Ask your child what tools he or she would use to measure different items in your home.

Practice Match to show the tool you can use to measure.

4. How much water does it hold?

5. How cold is it?

6. How heavy is it?

7. How long is it?

 Spiral Review and Test Prep

Choose the correct answer.

8. Which tool would you use to measure the height of the dinosaur?

- ○ scale
- ○ yardstick
- ○ thermometer
- ○ measuring cup

9. 50 + 25 = ☐

- ○ 65
- ○ 75
- ○ 80
- ○ 85

Name _____

Build a Dinosaur Diorama

Your class wants to make a dinosaur diorama.
What will you show?

You Decide!

1. Plan your diorama. Choose the box you will use. Make a list.

_____

_____

_____

_____

_____

_____

Workspace

2. Talk with a partner. What will you decide to include in your dinosaur diorama?

3. Draw a picture to show what your diorama will look like. Measure the length and width of the box. Label the parts.

Workspace

4. **What if** you decide to tell a measurement story about your diorama? What story could you tell?

_____

_____

_____

_____

Name _____

What Can a Footprint Tell Us?

Footprints and remains of things that lived long ago are called fossils.

What to do

1. Measure and cut a string 53 inches long to show a dinosaur footprint.

2. Then measure and cut a string the same length as your foot.

3. Find out how much longer the dinosaur footprint is than your footprint.

4. Now find out how many of your footprints make one dinosaur footprint.

McGraw-Hill School Division

Dinosaur Footprint

_____ inches

Your Footprint

_____ inches

Difference in Footprints

_____ inches

The largest dinosaur footprint ever found was about 53 inches long.

What did you find out?

5. How much longer was the dinosaur footprint than your footprint?

_____

6. How many of your footprints would it take to equal one dinosaur footprint?

_____

7. What can you tell about dinosaurs by looking at the size of their footprints?

_____

Journal

Want to do more?

Think of an animal. How much smaller or larger is your footprint than the animal's footprint? Find out.

Name _____

Use a centimeter ruler to measure.

1. about _____ centimeters

2. about _____ centimeters

Choose the better estimate.

3.

is less than 1 kilogram

is more than 1 kilogram

4.

is less than 1 kilogram

is more than 1 kilogram

5.

about 2 liters

about 200 liters

6.

about 4 liters

about 15 liters

Write each temperature.

7.

_____ °F

8.

_____ °F

9.

_____ °F

10.

_____ °F

Journal

What tool should you use to find the temperature outside? Why?

How Big?

- Work with a partner and another team.

- Find one object for each length on the chart.

- Write the name of the object on the chart.

- See which team can finish first.

Length	Found Object About This Length
about 2 inches	
about 7 inches	
about 4 feet	
about 2 centimeters	
about 7 centimeters	
about 18 centimeters	

Chapter Review

Name _____

Math Words

area
foot
inches
perimeter
yard
meter

Language and Math

Complete. Use a word from the list.

1. 12 _____ = 1 _____.

2. The distance around a shape is the _____.

3. _____ is measured in square units.

4. 100 centimeters = 1 _____.

Concepts and Skills

Use an inch ruler to measure.

6.

_____ inches

5.

_____ inches

7.

_____ inches

Choose the best estimate for each object.

8.

more than 1 cup

less than 1 cup

9.

more than 1 quart

less than 1 quart

10.

more than 1 pound

less than 1 pound

Find the perimeter.

11. _____ + _____ + _____ + _____

The perimeter is about _____ inches.

Color to show the number of square units.

12.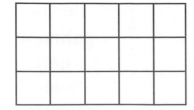

10 square units

13.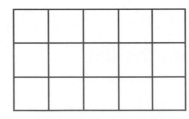

8 square units

14.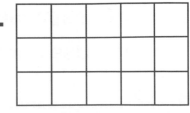

4 square units

Use a centimeter ruler to measure.

15. _____ centimeters

Choose the better estimate.

16.

more than 1 kilogram

less than 1 kilogram

17.

more than 1 liter

less than 1 liter

Write the temperature.

18. _____ °F

19. _____ °C

Problem Solving

20. Estimate the length. Then use a ruler to check.

Estimate _____ inches

Measure _____ inches

Name _____

Draw lines to connect the dots.

C

B•   •D

DINOSAUR
MUSEUM

H •   • G

A•   İ   Ḟ   •E

Use an inch ruler to measure. Complete each sentence.

1. From A to B is about _____ inches.

2. From A to E is about _____ inches.

3. From E to F is about _____ inches.

4. _____ to _____ is about the same length as _____ to _____.

Volume is the amount of space an object takes up.
Volume is measured in **cubic units**.

Use connecting cubes to build each figure.
Then write how many cubes you used.

1.

_____ cubes   _____ cubes   _____ cubes

2.

_____ cubes   _____ cubes   _____ cubes

Find the volume of each figure.

3.

_____ cubic units   _____ cubic units   _____ cubic units

4.

_____ cubic units   _____ cubic units   _____ cubic units

Name _____

Use a ruler to measure.

1.

about _____ inches

2.

about _____ centimeters

Choose the better estimate.

3.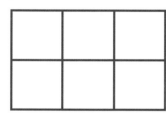

more than 1 quart

less than 1 quart

4.

less than 1 kilogram

more than 1 kilogram

5. Find the number of square units.

_____ square units

6. Find the perimeter.

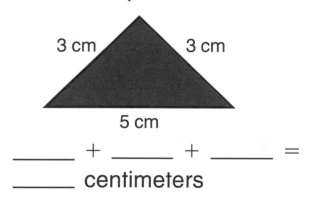

3 cm   3 cm

5 cm

_____ + _____ + _____ =

_____ centimeters

Write the temperature.

7. _____ °F

8. _____ °F

9. _____ °C

10. Estimate the length. Then use a ruler to check.

Estimate   Check

_____ inches   _____ inches

McGraw-Hill School Division

Performance Assessment

Remember, you can use a ruler to measure many different things.

Find two different objects to measure.
Measure them as many ways as you can.
Record each measurement and the tool you used.

Object 1	Object 2
measurement _____ _____ tool _____	measurement _____ _____ tool _____
measurement _____ _____ tool _____	measurement _____ _____ tool _____
measurement _____ _____ tool _____	measurement _____ _____ tool _____

Portfolio

You may want to put this page in your portfolio.

Name _____

Choose the correct answer.

Measurement and Geometry

1. How long does the show last?

Show begins 2:30 and ends 4:30

- ○ 1 hour
- ○ 2 hours
- ○ 3 hours
- ○ 4 hours

2. Which is the triangle?

- ○

- ○

- ○

- ○

3. Which has 4 corners?

- ○

- ○

- ○

- ○

Statistics, Data Analysis, and Probability

Use the graph to answer questions 4 and 5.

4. Who read the most books?

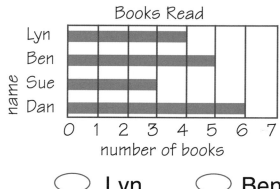

- ○ Lyn   ○ Ben
- ○ Sue   ○ Dan

5. How many books were read in all?

- ○ 6   ○ 18
- ○ 10   ○ 20

6. Look at this number pattern? What number would most likely come next?

3, 6, 9, 12, _____

- ○ 13   ○ 14
- ○ 15   ○ 16

Explain the pattern.

_____

Mathematical Reasoning

7. Draw a picture to model this problem.

There were 5 birds in a tree. 3 flew away. How many birds are left?

8. Don has 35¢. He spends 15¢. How much money does he have left?

- ⬭ 20¢
- ⬭ 25¢
- ⬭ 30¢
- ⬭ 45¢

9. Carl was in line behind Tim. Tim was in front of Allen. Allen was behind Larry.

List the order that the boys were in line.

- ⬭ Allen, Larry Carl, Tim
- ⬭ Tim Larry, Allen, Carl
- ⬭ Tim, Carl, Allen, Larry
- ⬭ Tim, Carl, Larry, Allen

Number Sense

10. Add.

$$27 + 18$$

- ⬭ 35
- ⬭ 38
- ⬭ 45
- ⬭ 48

11. Which number comes between?

29		31

- ⬭ 28
- ⬭ 30
- ⬭ 32
- ⬭ 33

12. Which digit is in the tens place?

54

- ⬭ 5
- ⬭ 4
- ⬭ 50
- ⬭ 40

theme

Cityscapes

Use the Data

Name the solid figures that you can find in the picture.

What You Will Learn

In this chapter you will learn how to:

- Identify 2-dimensional and 3-dimensional figures.

- Identify congruent and symmetrical figures.

- Use a pattern to solve problems.

Let's Tour the City!
Story by Ann McFadden
Illustrated by Peggy Togel

CITY TOURS

MATH AT HOME

Math Words

congruent
same size and shape

vertex →
edge →
face →

line of symmetry line
along which a figure can be
folded so that the two parts
match exactly

Dear Family,

In Chapter 10, I will learn about shapes and
ways to describe them.
Here are new vocabulary words and an activity
that we can do together.

Guess What?

- Place some objects that are solid
  shapes on a table.

Use

- Choose an object and give several
  clues to its identity. For example: I
  am round. I am flat. I have no
  corners. What am I?

- Take turns describing
  and identifying different
  objects.

Additional activities at
www.mhschool.com/math

Learn

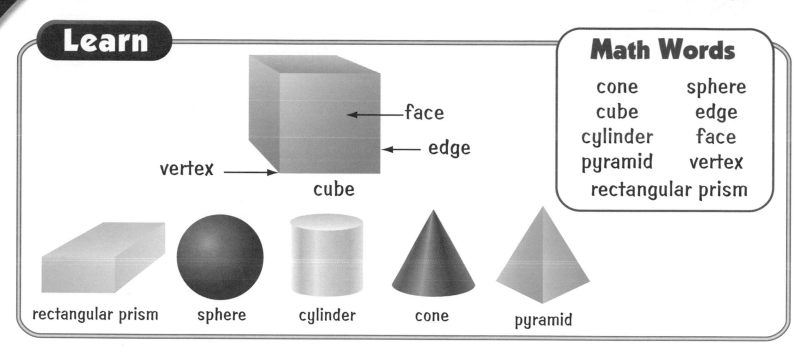

face

edge

vertex

cube

rectangular prism   sphere   cylinder   cone   pyramid

Math Words

cone	sphere
cube	edge
cylinder	face
pyramid	vertex
rectangular prism	

Try it

Circle the objects that have the same shape.

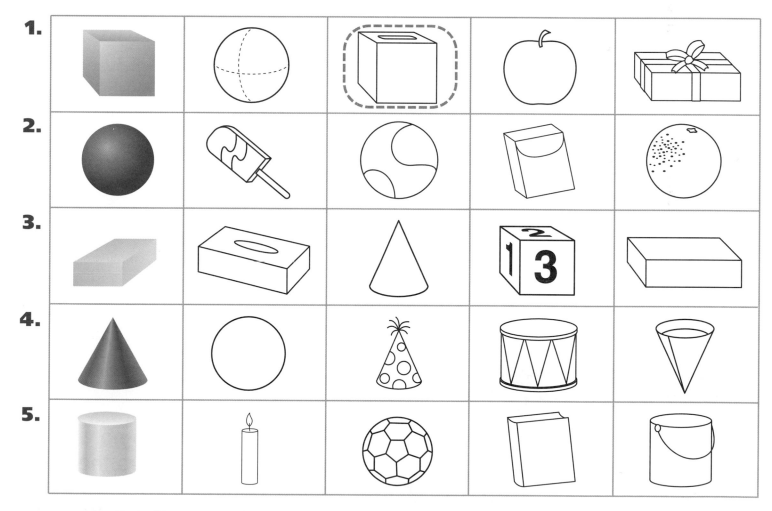

1.

2.

3.

4.

5.

Sum it Up How are the cube and rectangular prism the same?
How are they different?

 **Math at Home:** Your child learned about the attributes of solid figures.
Activity: Have your child identify the shapes of common objects in your home.

Practice

Circle the solid figure named. Write how many faces, vertices, and edges it has.

		Solid Figure	Faces	Vertices	Edges
6.	cube	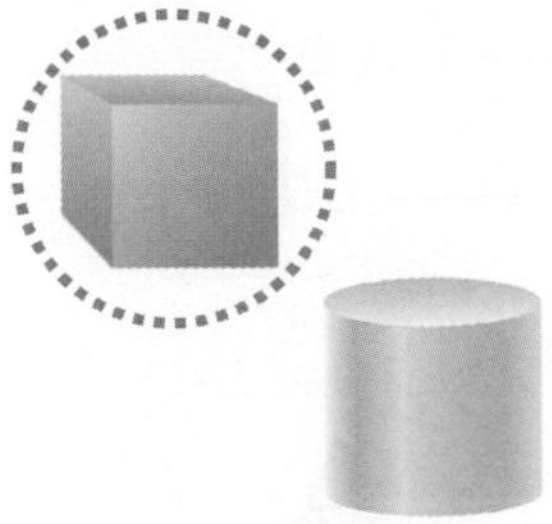	6	8	12
7.	rectangular prism	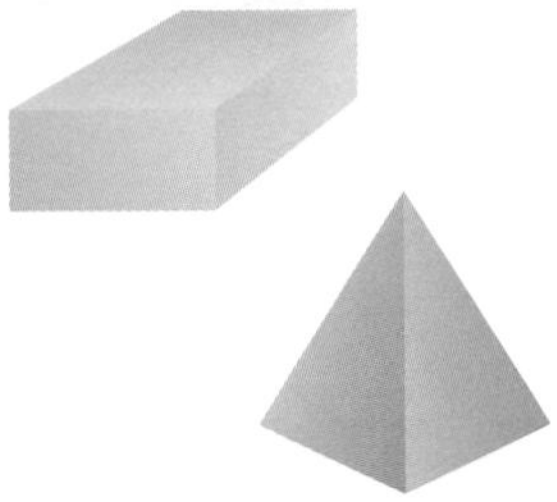	____	____	____
8.	pyramid	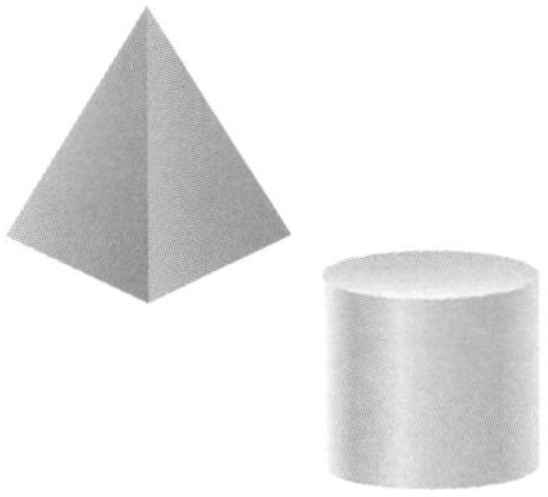	____	____	____
9.	rectangular prism	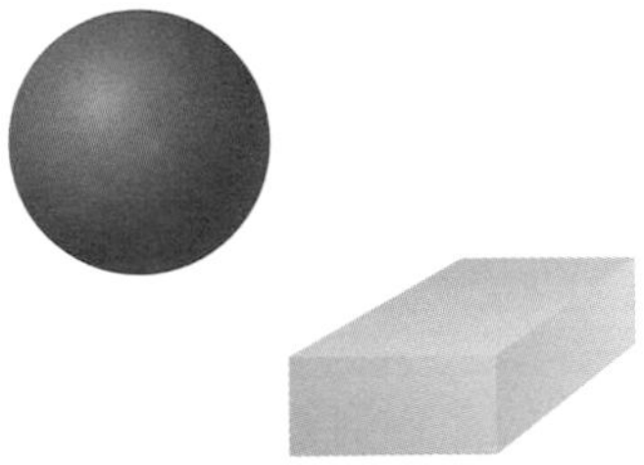	____	____	____

Problem Solving

10. I have 2 faces. I roll. Which solid figure am I?

____

11. I have 6 faces. I do not roll. All my edges are the same length. Which solid figure am I?

____

Name _____

Learn

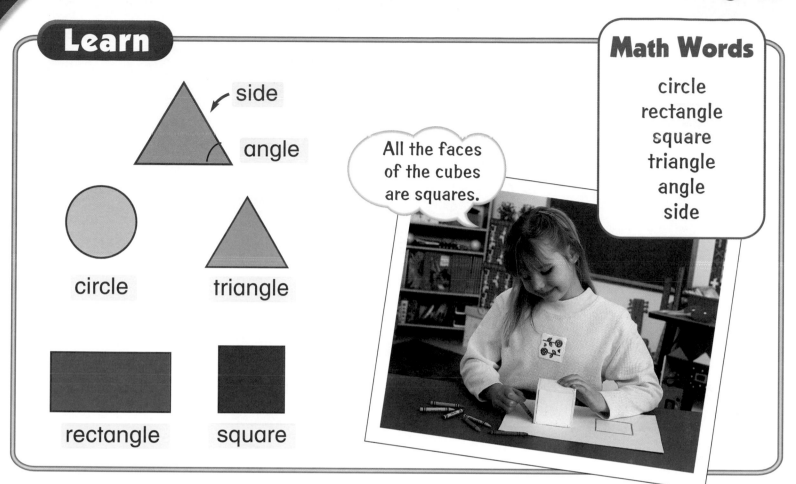

side

angle

circle

triangle

rectangle

square

All the faces of the cubes are squares.

Try it

Find objects with these shapes. Trace around one face. Circle the plane figure you made.

1.

2.

3.

4.

Sum it Up

You traced around the face of a solid figure to make a circle. What solid figure did you trace?

 Math at Home: Your child learned about the shapes of faces of solid figures.
Activity: Have your child trace the faces of a box and a can and name the faces he or she traced.

5.

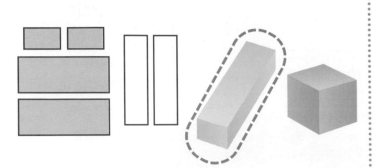

6.

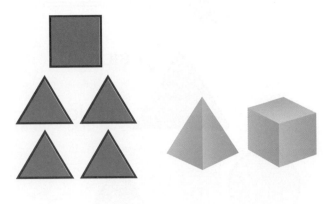

7.

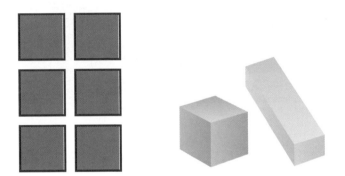

8.

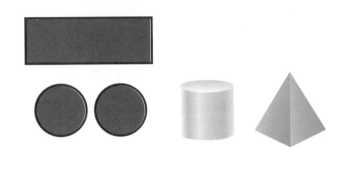

 Problem Solving

Mental Math

9. Amy and Tim have 30 blocks.
Amy builds this skyscraper.
Are there enough blocks for Tim
to build the same skyscraper?

_____

 Tell how you know.

_____

Learn

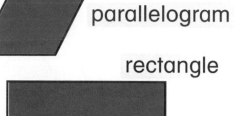

square

parallelogram

rectangle

Math Words

quadrilateral
parallelogram
pentagon

Each quadrilateral has
4 sides and 4 angles.

A pentagon has
5 sides and 5 angles.

Try it Tell how many sides and angles. Name each figure.

1.

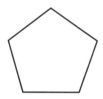

__5__ sides   __5__ angles

(pentagon)   square

2.

_____ sides   _____ angles

triangle parallelogram

3.

_____ sides   _____ angles

pentagon   square

4.

_____ sides   _____ angles

quadrilateral   triangle

Sum it Up How are squares, rectangles, and parallelograms alike?

_____

Math at Home: Your child learned about quadrilaterals and pentagons.
Activity: Have your child find examples of a parallelogram, a rectangle,
and a square.

three hundred fifty-seven **357**

Practice Color the figures named.
Then tell how many sides
and angles each has.

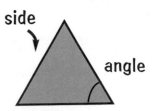

side

angle

5. quadrilateral

_____ sides   _____ angles

6. pentagon

_____ sides   _____ angles

7. parallelogram

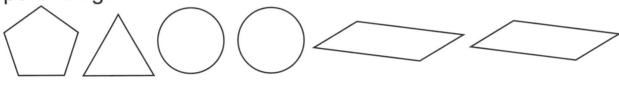

_____ sides   _____ angles

Spiral Review and Test Prep

Choose the correct answer. Fill-in the ◯.

8. Which figure has 5 sides and 5 corners?

9. What is the mode?

4, 3, 3, 2, 3

◯ 2
◯ 3
◯ 4
◯ 5

Learn

I can use 2 trapezoids to make a hexagon.

trapezoid

hexagon

Math Words

hexagon
trapezoid

A trapezoid has 4 sides.

A hexagon has 6 sides.

Try it   Use pattern blocks to make figures.
Then complete the chart.

Use these pattern blocks.	Make a new figure.	How many sides?	How many corners?	Name of new figure
1.		_6_	_6_	hexagon
2.		___	___	___

Sum it Up   Can you use squares to make squares, rectangles, and circles? Explain.

Math at Home: Your child combined figures to make new figures.
Activity: Have your child show you how to combine a square and a triangle to make a new figure.

three hundred fifty-nine   **359**

McGraw-Hill School Division

Use pattern blocks to make new figures. Complete the chart.

Use these blocks.	Make a new figure.	How many sides?	How many corners?	Name of new figure.
3.		_4_	_4_	<u>rectangle</u>
4.		___	___	___
5.		___	___	___

Problem Solving

Visual Thinking

6. Tommy made this figure using 3 pattern blocks. What were the figures that Tommy used?

_____

Draw lines to show how Tommy put the figures together.

Name _____

Make Decisions

Reading Skill   Making decisions can help you solve problems.

Mr. Green's class is building a city out of blocks. Amy builds a skyscraper. She uses different figures.

Build a skyscraper. Use different figures. Then answer the questions.

1. Which figure did you use to make the door? Why? _____

2. Which figures did you use to make the windows? Why? _____

3. Which figure did you use to make the roof? Why? _____

4. What does the front of the building look like? Draw the skyscraper.

 Math at Home: Your child learned to solve math problems by making decisions.
Activity: Have your child draw a picture of a building that is made of three different shapes put together.

Solve.

Dale builds a bridge. He uses different figures.

Build a bridge. Use different figures.
Then answer the questions.

1. Which figures did you use to make the end parts of the bridge?
   Why? _____

2. Which figures did you use to make road part of the bridge?
   Why? _____

3. Which figures did you use to make the top of the bridge?
   Why? _____

4. What does a side view of the bridge look like? Draw the bridge.

Name _____

Name each solid figure. Write how many faces, vertices, and edges.

	Solid Figure	Name	Faces	Vertices	Edges
1.		_____	_____	_____	_____
2.		_____	_____	_____	_____

What plane figure will you make if
you trace each solid figure? Circle it.
Then tell how many sides and angles.

3.

_____ sides _____ angles

4.

_____ sides _____ angles

Name the plane figure. Color the plane figure.

5. I have 4 sides and 4 corners.
Two of my sides are parallel.
The other two are also
parallel.
What am I?

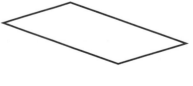

_____

Add or subtract.

1.

7	17	5	9	6	18
+ 8	− 8	+ 7	− 3	+ 8	− 9

Compare. Use <, >, or =.

2. 35 ◯ 47   59 ◯ 53   43 ◯ 28

Find the perimeter of each figure.

3.

4 in.

4 in.   4 in.

4 in.

_____ + _____ + _____ + _____

The perimeter is _____ inches.

5 in.   5 in.

7 in.

_____ + _____ + _____

The perimeter is _____ inches.

 TECHNOLOGY LINK

Put Shapes Together
- Use pattern blocks.
- Stamp out 2 triangles.
- Rotate one triangle 2 times.
- Put the triangles together.
- What shape is formed?

1. Stamp out other shapes and put them together. Tell the shape formed.

For more practice use Math Traveler™.

Name_____

Act It Out

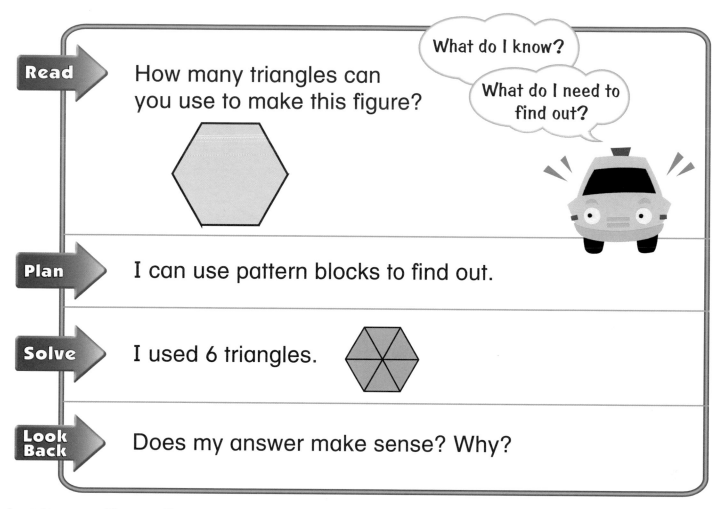

Read ▶ How many triangles can you use to make this figure?

What do I know?

What do I need to find out?

Plan ▶ I can use pattern blocks to find out.

Solve ▶ I used 6 triangles.

Look Back ▶ Does my answer make sense? Why?

Act it out. Draw lines.
Use triangles or squares to make each figure.

1.

_____ triangles

2.

_____ squares

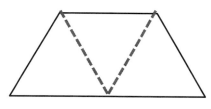 How does acting it out help you solve the problem?

 Math at Home: Your child used pattern blocks to solve problems.
Activity: Have your child show you how to make a rhombus using 2 triangles.

three hundred sixty-five **365**

Use triangles and squares to make each figure.

3.

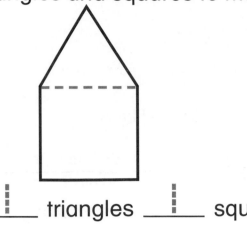

_____ triangles _____ squares

4.

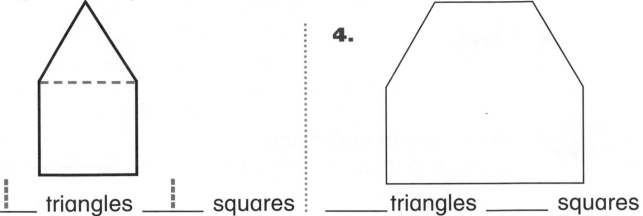

_____ triangles _____ squares

5.

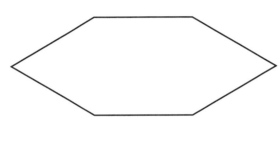

_____ triangles _____ squares

6.

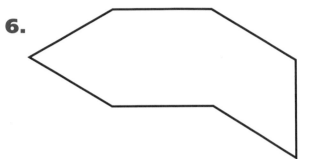

_____ triangles _____ squares

7.

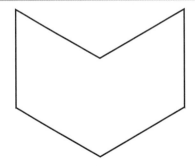

_____ triangles _____ squares

8.

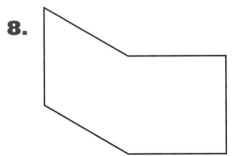

_____ triangles _____ squares

Critical Thinking **Journal**

9. Use these pattern blocks to make a building. Color it.

Workspace

Name _____

Learn

Congruent figures have the same size and shape.

Math Words

congruent

These triangles are congruent.

These squares are not congruent.

Try it

Color the two congruent figures.

1.

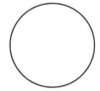 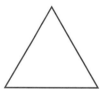   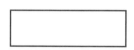

2.

 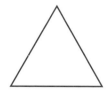 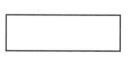 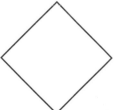

3.

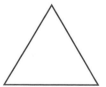 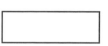 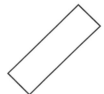

Sum it Up

When are two plane figures congruent?

 Math at Home: Your child learned about congruent figures.
Activity: Have your child show you two congruent figures.

three hundred sixty-seven **367**

Practice Draw a figure that is congruent.

4.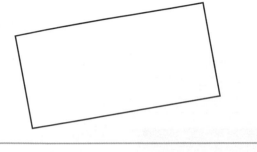

5.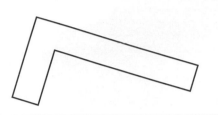

Color the two congruent figures.

6.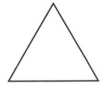

7.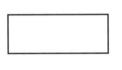

8.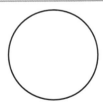

9.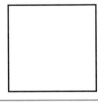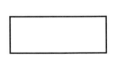

Algebra & functions Look for a pattern.
Write the missing numbers.

10. 15, 18, _____ 24, 27, _____ _____ ⋮ 29, 31, _____ _____ _____, 39, 41

11. 20, 25, _____ 35, _____ 45, _____ ⋮ 34, 38, _____ 46, 50, _____ 58

Learn

These figures have a line of symmetry. The two parts match exactly.

Math Words

line of symmetry

line of symmetry

These figures do not have a line of symmetry.

Try it

Cut out paper shapes.
Fold each shape to make a line of symmetry.
Draw the line in the fold.

1.

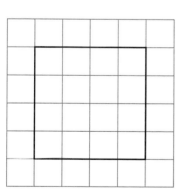

2.

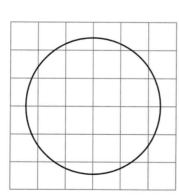

3.

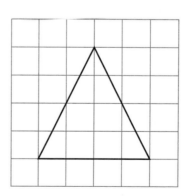

4.

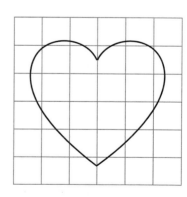

Sum it Up

What can you tell about a figure that has symmetry?

Math at Home: Your child learned about lines of symmetry.
Activity: Show your child a paper plate, a fork, and a knife.
Ask him or her if these shapes have lines of symmetry.

three hundred sixty-nine **369**

Draw a matching part for each figure.

5.

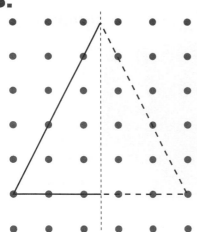

6.

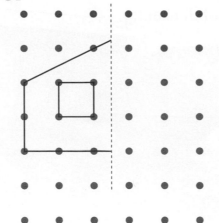

7.

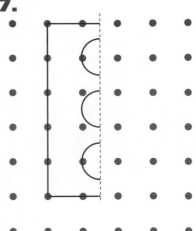

8.

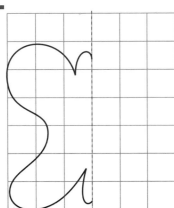

9.

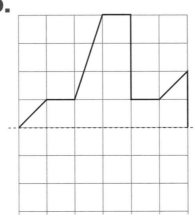

10.

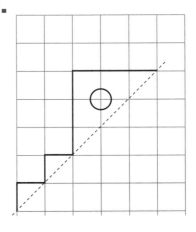

 Problem Solving

Visual Thinking

11. Which building has symmetry?
How do you know?

Name _____

Design a City Building

Your class wants to design a city building.
Which figures will you include?

Workspace

1. What structure will you design?

2. How many different figures will you use?
Make a list of your ideas.

_____

_____

_____

_____

_____

Tell which building you decided to build. _____
Draw a picture.

What if you wanted to design a restaurant? What kind of restaurant would you design? Which figures could you use? Explain.

_____

_____

_____

_____

Workspace

Name _____

What Kinds of Shapes Can You Find at Playgrounds?

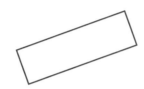

Playground structures are made up of many simple shapes.

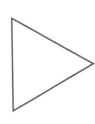

What to do

1. Use blocks. Make a playground structure.

2. Draw a picture of what you built.

3. Then tell about the structure.

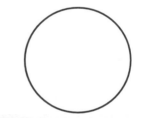

My Playground

Playground Structure

Kind of Shape	How Many?	What is structure for?

What did you find out?

1. What shapes did you use?

_____

2. Did you use different shaped blocks to show different parts of your structure?

Explain. _____

_____

3. What kind of shapes do you see at a playground near you?

_____

_____

Did You KNOW?

Architects make small models of structures they design.

Want to do more?

Make another structure. Use some different shapes.

Name_____

Shade the two shapes that are congruent.

1.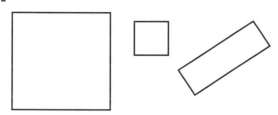

2.

Draw the matching part for each figure.

3.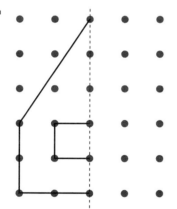

4.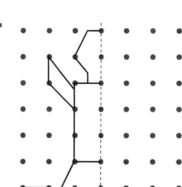

5.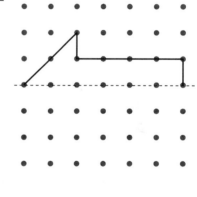

Use pattern blocks. How many triangles and squares does it take to make each figure?

6.

____ triangles ____ squares

7.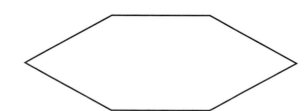

____ triangles ____ squares

Name_____

Build It

You Will Need

pattern blocks

- Take turns. Each player covers part of the house with one of the pattern blocks.

- Continue until the house is fully built.

- The winner is the player who places the last pattern block on the house.

Name _____

Language and Math

Choose the correct word to complete each sentence.

1. This figure is a _____.

2. This figure has a _____.

3. Two figures are _____ when they are the same size and shape.

Concepts and Skills

Name each solid figure.
Find how many faces, vertices, and edges.

Solid Figure	Name	Faces	Vertices	Edges
4.				
5.				
6.				

What figure would you make if you traced each solid?
Circle it. Name each figure.

7.

8.

9.

10.

_____   _____   _____   _____

Shade the figures that are congruent.

11.

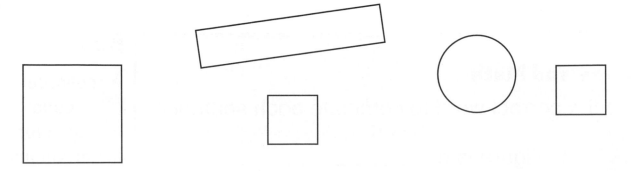

12.

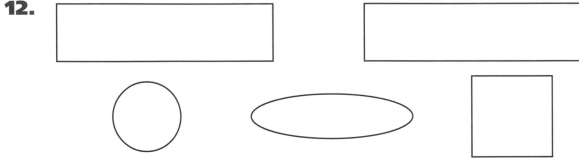

Draw a matching part for each figure.

13.

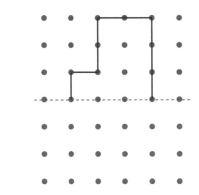   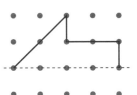

 Problem Solving

How many triangles or squares
does it take to make each figure?

14.

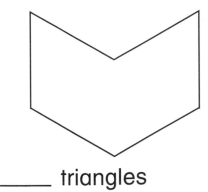

_____ triangles   _____ squares

Name _____

Shaping Up!

1. Find all of these plane figures. Color.

    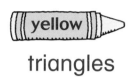

squares   rectangles   circles   hexagons   triangles

Slides, Flips, Turns

You can move these wagons in different ways.

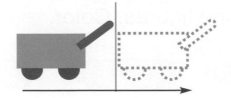
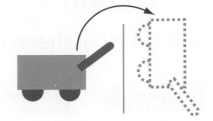

This is a flip.   This is a slide.   This is a turn.

Use a rectangle.
Put a yellow dot on the corner.

1. Put your rectangle on top of here. Slide it to here. Draw the dot.

   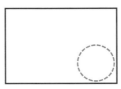

---

2. Put your rectangle on top of here.   Turn it to here. Draw the dot.   Put your rectangle on top of here.   Flip it. Draw the dot.

        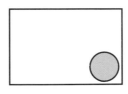   

Write the word that names the move.

3.

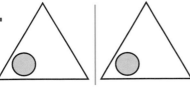   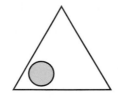   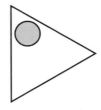   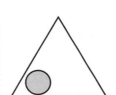   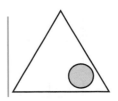

_____   _____   _____

Name _____

Circle each solid figure.

1. Cube

2. Cylinder

3. Pyramid

Circle each plane figure.

4. Triangle

5. Rectangle

6. Square

Color the two congruent figures.

7.

Draw a matching part for each figure.

8.

9.

Use pattern block triangles to make the figure.

10.

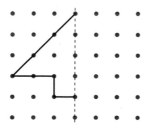

Tell what you know about each figure.

Figure	Name	How many edges?	How many corners?	What else do I know?
1.				
2.				
3.				
4.				

You may want to put this page in your portfolio.

Name_____

Choose the correct answer.

Algebra and Functions

1. Which is the turnaround fact for $8 + 9 = 17$?

- $8 + 8 = 16$
- $9 + 8 = 17$
- $9 + 9 = 18$
- $16 - 8 = 8$

2. Sonia had 16 crayons. She gave 2 crayons to her brother. How many crayons did she have left?

Which number sentence would you use to solve this problem?

- $16 - 2 = 14$
- $16 + 2 = 18$
- $20 + 2 = 22$
- $20 - 2 = 18$

3. Which number makes this sentence true?

$6 + \boxed{} = 14$

- 6
- 7
- 8
- 20

Measurement and Geometry

4. There are _____ days in 1 week.

- 5
- 7
- 12
- 60

5. Measure the length of the screw to the nearest centimeter.

- 1 centimeter
- 2 centimeters
- 3 centimeters
- 4 centimeters

6. Which is a cylinder?

-
-
-
-

Number Sense

7. Which are the even numbers?

○ 13, 14, 15
○ 16, 18, 20
○ 16, 17, 18
○ 18, 20, 21

8. I am a number between 70 and 90.
When you count by 10, you say my name.
What number am I?

○ 72
○ 75
○ 80
○ 85

9. Which number comes between 63 and 65?

○ 62
○ 64
○ 66
○ 70

Mathematical Reasoning

10. Which unit would you use to measure how far a park is from your house?

○ foot
○ miles
○ pound
○ minute

11. Joey says the time shown is 1:30. Is he correct? Tell how you know.

12. Bobby says the answer to this problem is 30 crackers. Is he correct? Tell how you know.

Todd had 17 crackers.
He ate 3 of them.
How many crackers were left?

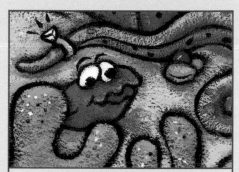

Use the Data

What fraction stories can you tell about this picture?

What You Will Learn

In this chapter you will learn how to:

- Identify fractional parts of a group and of a whole.

- Identify likely, unlikely, and improbable events.

- Use logical reasoning to solve problems.

Math at Home

Dear Family,

In Chapter 11, I will learn fractions and probabilities. I will learn about halves, fourths, eighths, thirds, sixths, and twelfths. Here are new vocabulary words and an activity that we can do together.

Penny Fractions

Math Words

fraction

A number that names part of a whole or a group.

halves
$\frac{1}{2}$

thirds
$\frac{1}{3}$

fourths
$\frac{1}{4}$

eighths
$\frac{1}{8}$

twelfths
$\frac{1}{12}$

- Take 12 pennies and spread them on the table.

- Ask your child to count how many pennies there are.

- Then ask him or her to separate them into 2 equal groups.

- Ask how many pennies there are in each group.

- Ask what fraction each share represents.

- Repeat the activity with 3, 4, and 6 equal groups.

use

12 pennies

Additional activities at www.mhschool.com/math

Name _____

Learn

You can use a fraction to name equal parts.

2 equal parts
halves

  $\frac{1}{2}$

1 of 2 equal parts
one half

4 equal parts
fourths

  $\frac{1}{4}$

1 of 4 equal parts
one fourth

8 equal parts
eighths

  $\frac{1}{8}$

1 of 8 equal parts
one eighth

Math Words

fraction
one half
one fourth
one eighth
halves
fourths
eighths

Fractions show parts of a whole.

Try it   Write the fraction for the shaded part.

1.

 $\frac{1}{2}$

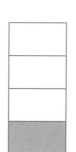 _____

 _____

   How could you show $\frac{1}{2}$ of a square? $\frac{1}{4}$ of a square? $\frac{1}{8}$ of a square?

McGraw-Hill School Division

Math at Home: Your child learned about fractions as parts of a region.
Activity: Give your child a sheet of paper. Ask him or her to color one-fourth green. Repeat with another sheet and ask him or her to color one-eighth red.

Practice
Color one part of each figure.
Then write the fraction for the part.

2. 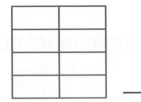 $\frac{1}{2}$ 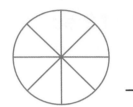 _____ 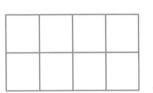 _____

3. 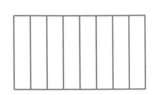 _____  _____ 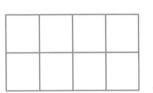 _____

4.  _____  _____ 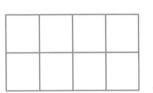 _____

Answer each question. Write yes or no.

5.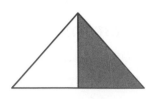

Is this $\frac{1}{2}$?   Is this $\frac{1}{4}$?   Is this $\frac{1}{8}$?

_____   _____   _____

Critical Thinking Journal

6. Would you rather have $\frac{1}{2}$ of the box of raisins or $\frac{1}{4}$ of the box of raisins? Explain.

_____

Name _____

Learn

When the parts are equal, you can write a fraction.

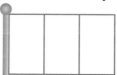

3 equal parts
thirds

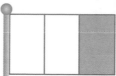 $\frac{1}{3}$

1 of 3 equal parts
one third

6 equal parts
sixths

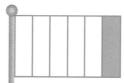 $\frac{1}{6}$

1 of 6 equal parts
one sixth

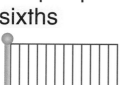

12 equal parts
twelfths

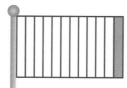 $\frac{1}{12}$

1 of 12 equal parts
one twelfth

Try it   Write the fraction for the shaded part.

1.

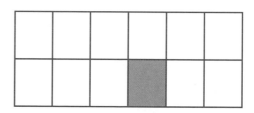

$\frac{1}{3}$ _____   _____

Sum it Up

What does $\frac{1}{6}$ mean?

 Math at Home: Your child learned more about fractions of a region.
Activity: Have your child show you how to divide a sheet of paper into 6 equal parts.

Color one part of each figure.
Then write the fraction for the part.

2. 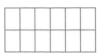 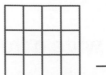 _____ 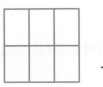 _____

3. _____   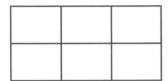 _____   _____

Write the word and the fraction for each word.

4.

Each part is one _____, or _____.

5.

Each part is one _____, or _____.

6.

Each part is one _____, or _____.

Spiral Review and Test Prep

Choose the correct answer. Fill-in the ◯.

7. Which picture shows that $\frac{1}{4}$ of the cracker has been eaten?

◯    ◯ 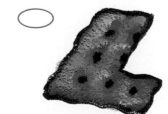   ◯ 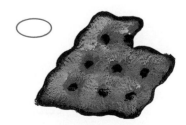   ◯

8. $25 + 25 + 25 = \boxed{}$

◯ 50   ◯ 70   ◯ 75   ◯ 80

Learn

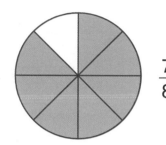

$\frac{4}{4}$ is the same as I.

These fractions name the number of equal parts that are shaded.

3 of 4 equal parts

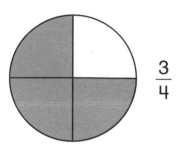

$\frac{3}{4}$

three fourths

7 of 8 equal parts

$\frac{7}{8}$

seven eighths

A fraction can name from, all, or none equal part of a whole.

Try it   Write the fraction for the part that is shaded.

1.

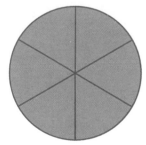

$\frac{2}{3}$

___

___

Sum it Up  How do you show $\frac{7}{12}$ of a rectangle?

🏠 **Math at Home:** Your child learned about fractions.
Activity: Have your child draw a rectangle and draw lines to show
6 equal parts. Ask your child to color 5/6.

Practice

Write the fraction for the part that is shaded.

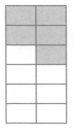

> Remember, fractions can name more than one equal part.

2. 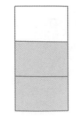 $\dfrac{2}{3}$ _____   _____

3. _____ 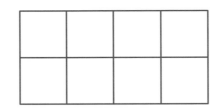 _____ 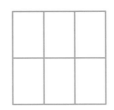 _____

Color to show the fraction.

4.  $\dfrac{2}{4}$ _____ $\dfrac{3}{8}$ _____ $\dfrac{5}{6}$

5. 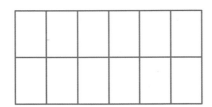 $\dfrac{7}{12}$ 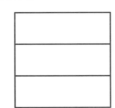 $\dfrac{2}{3}$  $\dfrac{7}{8}$

 Problem Solving

Look at the picture. Solve.

6. The children made kites.
What fraction of the kite is purple? _____

How do you know?

_____

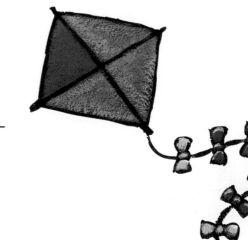

392 three hundred ninety-two

Name _____   **Compare Fractions**

Learn

Divide the whole into 2 equal parts.

How many parts?

Name the fraction for the shaded part.

$$\frac{1}{2}$$

Is the fraction greater than or less than the whole?

less than

Divide each part into 2 equal parts.

How many parts?

What fraction is the shaded part?

$$\frac{1}{4}$$

Is the shaded part greater than or less than $\frac{1}{2}$?

less than

Divide again.

How many parts?

What fraction is the shaded part?

$$\frac{1}{8}$$

Is the shaded part greater than or less than $\frac{1}{4}$?

less than

Try it

Circle the fraction that is greater.

Each time you divide the parts, each part becomes smaller.

1.

      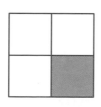

$\frac{1}{2}$ $\frac{1}{4}$

2.

   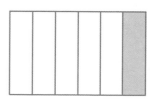

$\frac{1}{3}$ $\frac{1}{6}$

Sum it Up

Which fraction is less $\frac{1}{4}$ or $\frac{1}{6}$? Explain.

Math at Home: Your child compared fractions.
Activity: Divide one sheet of paper into 2 equal parts. Divide another sheet of paper into 4 equal parts. Ask your child what fraction each part shows. Ask him or her to tell you which fraction is greater.

three hundred ninety-three **393**

Practice   Circle the fraction that is greater.

3.

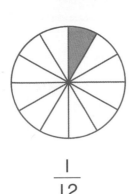

$\dfrac{1}{12}$ $\left(\dfrac{1}{6}\right)$

4.

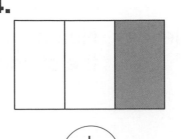

$\left(\dfrac{1}{3}\right)$ $\dfrac{1}{8}$

5.

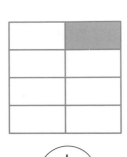

$\left(\dfrac{1}{8}\right)$ $\dfrac{1}{12}$

6.

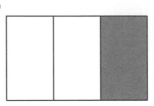

$\left(\dfrac{1}{3}\right)$ $\dfrac{1}{6}$

Compare the fractions. Use <, >, or =.

7. $\dfrac{1}{2} \bigcirc \dfrac{1}{4}$ $\dfrac{1}{2} \bigcirc \dfrac{1}{6}$ $\dfrac{1}{2} \bigcirc \dfrac{1}{8}$

8. $\dfrac{1}{4} \bigcirc \dfrac{1}{2}$ $\dfrac{1}{4} \bigcirc \dfrac{1}{6}$ $\dfrac{1}{4} \bigcirc \dfrac{1}{8}$

9. $\dfrac{1}{6} \bigcirc \dfrac{1}{2}$ $\dfrac{1}{6} \bigcirc \dfrac{1}{4}$ $\dfrac{1}{6} \bigcirc \dfrac{1}{8}$

Critical Thinking   Journal

10. Write each fraction.
Then put the fractions in order from greatest to least.

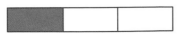   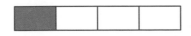   

_____   _____   _____

_____

Name _____

Making Predictions

Reading Skill You can read a story to make predictions.

Amy and Lucas had a Celebrate the Sea! celebration.

Amy ate $\frac{1}{2}$ of a pizza.

Lucas ate $\frac{1}{3}$ of a pizza.

Each said, "I ate more."

Solve.

1. How much of the cheese pizza is left? _____

2. How much of the veggie pizza is left? _____

3. Which pizza do you think Amy ate? _____

4. Is it possible that Lucas ate more than Amy? Explain.

_____

Math at Home: Your child read a story to make predictions.
Activity: Give your child a cutout circle and have him or her divide the paper into fourths and then color 3/4. Ask how much of the circle is left to color.

three hundred ninety-five **395**

McGraw-Hill School Division

Solve.

George and Kelly ate sandwiches at the party. George ate $\frac{1}{4}$ of a sandwich. Kelly ate $\frac{1}{6}$ of a sandwich.

Solve.

5. How much of the turkey sandwich is left? _____

6. How much of the veggie sandwich is left? _____

7. Which sandwich do you think George ate? _____

8. Is it possible that Kelly ate more than George? Explain.

_____

9. If Kelly ate another sixth of her sandwich, how much would be left? _____

10. If George ate another fourth of his sandwich, how much would be left? _____

Name _____

Color one part of each figure.
Then write the fraction for the part.

1.  _____   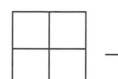 _____    _____

Write the fraction for the part that is shaded.

2. _____   _____   _____

3. _____   _____   _____

Color to show the fraction.

4.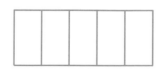

$\frac{1}{2}$ $\frac{3}{3}$ $\frac{1}{5}$

Compare the fractions. Use <, >, or =.

5. $\frac{1}{2}$ ◯ $\frac{1}{4}$ $\frac{1}{3}$ ◯ $\frac{1}{6}$ $\frac{1}{8}$ ◯ $\frac{1}{4}$ $\frac{1}{2}$ ◯ $\frac{2}{2}$

6. $\frac{1}{4}$ ◯ $\frac{1}{2}$ $\frac{1}{5}$ ◯ $\frac{1}{6}$ $\frac{1}{4}$ ◯ $\frac{1}{3}$ $\frac{1}{3}$ ◯ $\frac{2}{2}$

Add or subtract.

1.

23	36	30	38	99	49
+ 70	− 7	− 26	+ 3	− 9	+ 8

2.

76	59	80	72	63	45
− 21	+ 12	− 9	+ 12	+ 17	− 45

Write each word.

3. 28 _____   74 _____

TECHNOLOGY LINK

Compare Fractions
- Use fractions. Click on the rectangle.
- Click the up arrow 7 times.
- Color 3 parts blue.

- What fraction names the

  blue part? _____

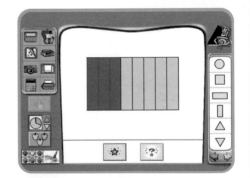

- What fraction names the gray part? _____

- Which fraction is greater? _____

   1. Show two other fractions. Tell which fraction is greater.

For more practice use Math Traveler™.

Name _____

Draw a Picture

What do I know?

Read The Sea Animal mural is divided into 8 equal parts. Dana draws a picture on the mural. She uses 2 parts.

What fraction of the mural does Dana use?

What do I need to find out?

Plan First I can draw a picture of the mural.

Then I can draw lines to divide the mural into 8 equal parts.

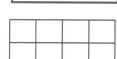

Solve There are 8 equal parts, so Dana uses

$\dfrac{2}{8}$ of the whole mural.

Look Back Does my answer make sense? Why?

Draw a picture. Solve.

1. Doug's card has 6 parts that are the same size. He colors 1 part green. What fraction of the card does Doug cover?

_____

How does drawing a picture help you solve the problem?

Math at Home: Your child drew a picture to answer questions relating to fractions.
Activity: Fold a sheet of paper in half twice. Have your child color one section. Ask what fraction of the paper he or she colored.

Practice   Draw a picture. Solve.

2. Jackie makes a big banner for the school party. It has 5 equal parts. She paints 4 parts of it blue. She paints 1 part yellow.

What fraction of the banner is yellow?

_____

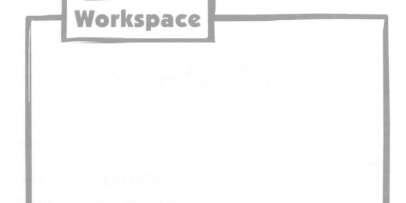
Workspace

___

3. Francine paints a big hexagon. Five of the six equal sections are blue. One section is red.

What fraction of the hexagon is red?

_____

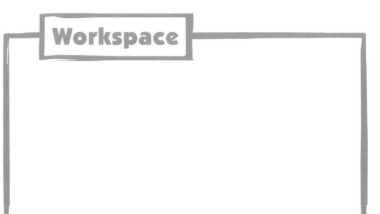
Workspace

___

Problem Solving

Critical Thinking

4. Look at the picture. Use fractions. Write a problem.

_____

_____

Learn

What fraction of the counters are red?

I of the counters is red.
There are 4 counters in all.

So $\dfrac{1}{4}$ of the counters are red.

How many counters are red? ___ $\dfrac{1}{4}$

How many counters in all? ___

$\dfrac{1}{4}$ of the counters are red.

$\dfrac{1}{4}$

1 ← number of red counters

4 ← total number of counters

Try it   Circle the fractions for the part that is yellow.

1.
$\dfrac{2}{6}$ $\dfrac{5}{6}$ $\dfrac{6}{6}$

2.

$\dfrac{1}{8}$ $\dfrac{3}{8}$ $\dfrac{4}{8}$

Sum it Up How do you know which number belongs on the top of a fraction? on the bottom?

 Math at Home: Your child practiced finding fractions of groups.
Activity: Give your child 5 red beans. Ask your child to make a fraction that shows I/5.

3. $\dfrac{4}{6}$

4. $\dfrac{4}{8}$

5. $\dfrac{6}{12}$

Look at the picture. Write the fraction.

6. What fraction of the fish are green?

$\boxed{}$ ⟵ number of green fish

$\boxed{}$ ⟵ total number of fish

Critical Thinking

7. Look at the picture.

Kim said that $\dfrac{1}{2}$ of the fish are yellow.

Larry said that $\dfrac{3}{6}$ of the fish are yellow.

Who is right? _____

Explain how you know. _____

Name _____

Learn

What fraction of the fish are blue?

 $\frac{1}{2}$ is the same as $\frac{3}{6}$.

group of 6 fish   2 equal parts $\frac{1}{2}$ of the fish are blue.

$\frac{1}{2}$ of the fish are blue.

Try it   Color to show the fraction.

1. $\frac{3}{8}$ of the starfish are blue.

$\frac{1}{3}$ of the tubfish are orange.

$\frac{1}{2}$ of the sea lions are brown.

2. $\frac{1}{4}$ of the striped fish are red.

$\frac{2}{3}$ of the seals are brown.

$\frac{3}{3}$ of the sea otters are yellow.

Sum it Up

What number do you write for the bottom number of a fraction?

 **Math at Home:** Your child practiced finding fractions of groups.
Activity: Put a mix of soup spoons and teaspoons on the table. Ask your child to tell you what fraction of the spoons are teaspoons.

Color to show the fraction.

3. $\frac{3}{4}$ of the fish are purple.

$\frac{5}{8}$ of the fish are blue.

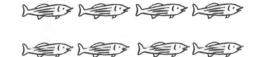

$\frac{1}{2}$ of the fish are green.

4. $\frac{1}{3}$ of the fish are red.

$\frac{1}{4}$ of the fish are pink.

$\frac{2}{3}$ of the fish are yellow.

Look at the picture. Answer each question.
Write the fraction for each part.

5. orange fish   yellow fish   green fish   purple fish

_____   _____   _____   _____

Draw a picture to solve the problem.

6. Andy has 3 red shells and 5 yellow shells. How many shells does he have?

_____ shells

What fraction of the shells are red?

_____

Workspace

Name _____

Learn

What color fish do you think he will pick, orange or purple?

Count the fish: ___4___ purple ___2___ orange.

Which color fish is he most likely to pick?

___purple___

Which color fish is he least likely to pick? ___orange___

Try it

Put cubes in a paper bag.

Which color are you most likely to pick?

Which color are you least likely to pick?

Pick one cube without looking.

Color each cube.

Bag	Most Likely	Least Likely	Your Pick
	◻	◻	◻
	◻	◻	◻
	◻	◻	◻

 **Sum it Up**

If there are more pink shells than yellow shells on the table, and you pick one without looking, which color are you most likely to pick? Explain why.

 Math at Home: Your child found the most likely and least likely outcomes.
Activity: Have your child put four dimes and two pennies in a paper bag. Ask your child to tell you which he or she is most likely to pick from the bag without looking. Have him or her then pick a coin from the bag without looking.

four hundred five

McGraw-Hill School Division

Put cubes in a paper bag.

Which color are you most likely to pick?

Which color are you least likely to pick?

Pick one cube without looking.

Color each cube.

Bag	Most Likely	Least Likely	Your Pick

Would you rather have $\frac{1}{2}$ of the box

of fish crackers or $\frac{3}{6}$ of the fish crackers?

_____

_____

Name _____   **Make Predictions**

Learn

You can make a prediction that something will happen.

I predict that I will spin green again.

That is the third green. We only have I red.

Try it   Use red and green to color a spinner. Answer each question.

1. What color do you predict you will spin more often?

   _____

2. Why do you think so? _____

red	green

3. Predict. If you spin the spinner 10 times, how many times will you get red?

   _____

   Spin the spinner 10 times.
   Record each spin.

4. How many times did the spinner land on green? _____

Sum it Up   How does knowing how something happened before help you predict how it is going to happen next?

Math at Home: Your child predicted spinner outcomes.
Activity: Toss a coin and have your child record how it landed. Repeat five times. Ask your child to predict how it will land next.

McGraw-Hill School Division

A bag of toy fish has 8 red, 4 blue, 4 yellow, and 4 green. Each child picks a toy fish from the bag then puts it back.

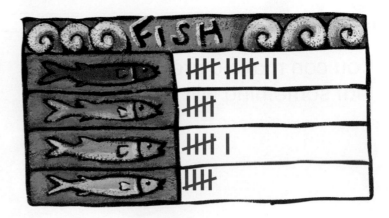

5. Which fish was picked most often? _____red_____

6. Barbara picks next. Is it very likely that she will pick a yellow fish?

_____

7. What do you predict Barbara will pick? _____

8. Why? _____

9. Mario made a spinner. These are his spins.

Blue: 6 times

Red: 1 time

Green: 0 times

White: 1 time

Which spinner do you think is Mario's? Circle it.

Tell how you know.

_____

Name _____

Sea Animal Game

You want to plan a game.
Look at the game board.
Use fractions in your rules.

Decide what rules you want to use.
Make a list.

Workspace

1. How did you decide what rules to use? _____

   How did you use fractions in your rules? _____

2. Draw a picture to show how
   you play the Sea Animal Game.

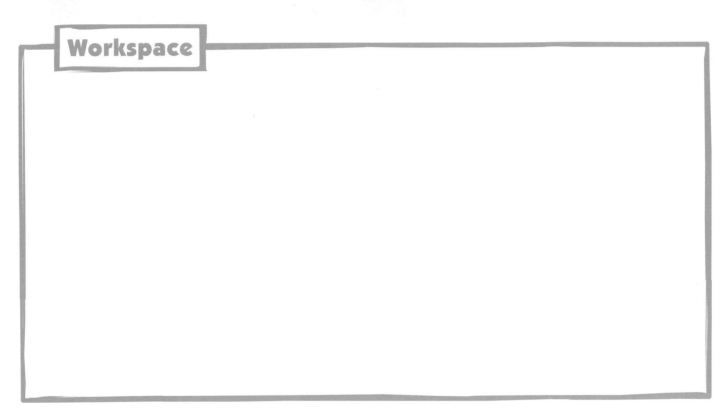

Workspace

3. **What if** you wanted to change the rules
   to make them harder?

   How would you change them?

   _____

   _____

Name _____

Why Is Water Important?

All living things need clean water to survive.

What to do

1. Fill a jar with water and put in a spoonful of oil. Put on the cap.

2. Now shake the jar. What happens?

3. Scoop out some oil with a spoon. What happens?

4. Add some soap, and shake the jar again. What happens?

5. Scoop out some of of the oil and soap.

You Will Need

small jar with lid

plastic spoons

liquid soap

vegetable oil

What happened?	When oil was added?	When soap was added?
_____	_____	_____
_____	_____	_____

6. Is it easier or harder to remove oil from water when soap is added?

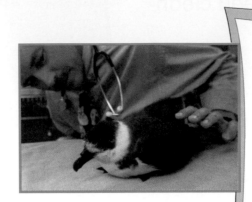

Did You KNOW?

Veterinarians use mild soap to clean oil off birds' feathers.

What did you find out?

7. Where was the oil before you shook the jar?

8. What happened to the oil after you shook the jar? Explain.

_____

_____

9. What happened to the oil when you added soap? Was it easier or harder to take out the oil? Explain.

_____

_____

Journal

Want to do more?

What if you put a spoonful of paint in water?
What do you think would happen? Explain.

Name _____

Color to show each fraction.

1. $\frac{3}{4}$ $\frac{1}{6}$ of the starfish are yellow. $\frac{5}{8}$ of the starfish are orange.

Look at the fish. What are you most likely to pick?
What are you least likely to pick? Color the fish.

2.    most likely   least likely

Color to show the fraction.

3. $\frac{2}{8}$ of the starfish are purple $\frac{1}{2}$ of the tubfish are orange $\frac{1}{3}$ of the seals are brown

4. Jerry made a tally chart to record spins on a spinner. This is what he has spun so far. What is he most likely to spin next?

_____

Spins	
red	blue
卌 III	II

5. 3 parts of a sign are blue. I part is green.

What fraction of the sign is green? _____

What fraction is blue? _____

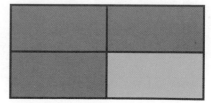

Name _____

Name a Fraction

- You and your partner take turns.

- Put the cubes in the bag. Then shake the bag.

- Take 4 cubes from the bag.

- Circle the fraction that names the red part.

- Put the cubes back in the bag.

- Play again. Take turns.

Play until one player circles all the fractions on his or her chart.

You Will Need

4 yellow cubes
4 red cubes
a paper bag
pencils
game cards

Player 1	
$\frac{1}{4}$	$\frac{2}{4}$
$\frac{3}{4}$	$\frac{4}{4}$

Player 2	
$\frac{1}{4}$	$\frac{2}{4}$
$\frac{3}{4}$	$\frac{4}{4}$

Name _____

Language and Math

Complete. Use a word from the list.

Math Words

half
third
fourth
eighth
fraction

1. A _____ shows equal parts.

2. A sandwich has 3 equal parts. Each part is one

   _____.

3. If $\frac{3}{4}$ of a circle is blue and the rest is yellow,

   one-_____ is yellow.

4. If a circle has 8 equal parts, each part is one

   _____.

5.

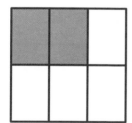

_____   _____   _____

6.

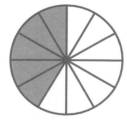
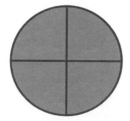

_____   _____   _____

Compare the fractions. Use <, >, or =.

7. $\dfrac{1}{2}$ ◯ $\dfrac{1}{8}$ $\dfrac{1}{6}$ ◯ $\dfrac{1}{3}$ $\dfrac{1}{4}$ ◯ $\dfrac{1}{8}$

Color to show each fraction.

8. $\dfrac{1}{5}$ $\dfrac{3}{4}$ $\dfrac{7}{12}$

You pick a fish from the bag without looking.
Which one are you the most likely to get?
Which are you least likely to get? Color the fish.

9.   most likely   least likely

Look at the table at the right.

10. What color are you most likely
to spin? _____

11. What color are you least likely to spin?

_____

 **Problem Solving**

Use logical reasoning. Solve.

12. Lydia's mural has 8 sections
She colors one section blue.
What fraction of the mural
does Lydia color blue?

Workspace

_____

Name _____

Write the fraction that names the shaded part.

1.  _____    _____   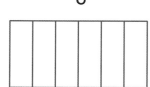 _____

Color to show each fraction.

2. $\dfrac{5}{8}$ $\dfrac{1}{3}$ $\dfrac{2}{6}$

Circle the fraction that shows how many are red.

3.

$\dfrac{1}{6}$ $\dfrac{2}{6}$ $\dfrac{3}{6}$ $\dfrac{2}{12}$ $\dfrac{3}{12}$ $\dfrac{4}{12}$ $\dfrac{1}{8}$ $\dfrac{1}{3}$ $\dfrac{3}{8}$

Color to show each fraction.

4. $\dfrac{3}{4}$ of the fish are blue. $\dfrac{1}{6}$ of the fish are yellow. $\dfrac{2}{8}$ of the fish are purple.

four hundred seventeen

Fractions can describe parts of a whole or parts of a group.

You have 8 counters. If $\frac{1}{2}$ of the counters are blue, how many counters are blue?

Separate the counters into 2 equal parts.

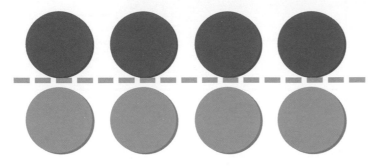

There are 4 counters in each equal part. $\frac{1}{2}$ of 8 is 4.

Draw pictures to find the part.

1. $\frac{1}{3}$ of 9 is _____.

2. $\frac{1}{4}$ of 8 is _____.

3. $\frac{1}{2}$ of 6 is _____.

4. $\frac{1}{5}$ of 10 is _____.

Name _____

Write the fraction for each green part.

1.

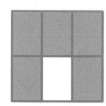 

_____   _____

Color to show the fraction.

2.

Color $\frac{3}{8}$ of the fish orange.   Color $\frac{1}{6}$ red.

Show the most likely and least likely outcomes.

3. Look at the paper bag. Which color are you most likely to pick? Which color are you least likely to pick? Color each cube.

Bag	Most Likely	Least Likely
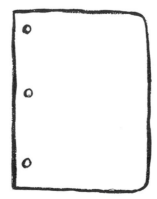		

---

4. Compare the fractions. Use <, >, or =.

$\frac{1}{6}$ ◯ $\frac{1}{4}$ $\frac{1}{3}$ ◯ $\frac{2}{2}$ $\frac{1}{2}$ ◯ $\frac{1}{4}$ $\frac{1}{8}$ ◯ $\frac{1}{4}$

💡 **Problem Solving**

5. Tony folded a paper into 8 parts. He colored 5 of the parts. What fraction of the paper did he color?

_____

Color to show the fractions.

1. Color $\frac{1}{2}$ blue.

Color $\frac{1}{4}$ green.

How many parts of the square are not colored?

_____

2. Color $\frac{1}{2}$ red.

Color $\frac{1}{4}$ yellow.

Color $\frac{1}{8}$ blue.

Color $\frac{1}{8}$ green.

How many parts of the square are not colored?

_____

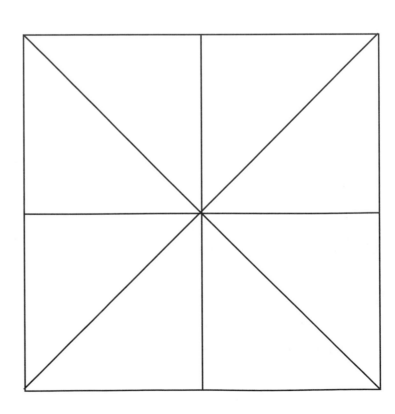

You may want to put this page in your portfolio.

...d twenty

Name _____

Choose the correct answer.

Number Sense

1. 52
    + 28

    ○ 70
    ○ 71
    ○ 80
    ○ 90

2. Which statement is true?

    ○ 48 = 48
    ○ 48 < 40
    ○ 48 > 60
    ○ 48 = 40

3. Which number has 9 in the tens place?

    ○ 39
    ○ 33
    ○ 97
    ○ 23

Algebra and Functions

4. Which number completes the subtraction sentence?

    $\boxed{} - 20 = 50$

    ○ 60
    ○ 70
    ○ 80
    ○ 90

5. Which number completes the addition sentence?

    $\boxed{} + 40 = 40$

    ○ 80
    ○ 40
    ○ 1
    ○ 0

6. Which number make this sentence true?

    half hour = $\boxed{}$ minutes

    ○ 10
    ○ 15
    ○ 30
    ○ 60

Mathematical Reasoning

7. Which unit would you use to measure the weight of a cat?

- ◯ foot
- ◯ minute
- ◯ miles
- ◯ pound

8. Leah had $2.35.
She lost one coin.
Now she has $2.25.
Which coin did she lose?

- ◯ a penny
- ◯ a nickel
- ◯ a dime
- ◯ a quarter

9. Julie had 30 green fish.
She gave Toby 16 green fish. How many fish did Julie still have?

Luis says the answer to this problem is 14 fish.
Is he correct?
Tell how you know.

_____

Statistics, Data Analysis and Probability

10. What is the range of the following set of numbers?
44, 11, 20, 25, 44

- ◯ 44
- ◯ 25
- ◯ 20
- ◯ 33

11. Look at the graph.
Which holiday got less than 3 votes?

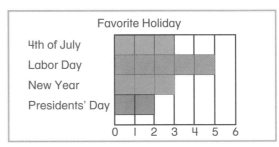

- ◯ 4th of July
- ◯ Labor Day
- ◯ New Year
- ◯ Presidents' Day

12. What does the word mode mean?

- ◯ the largest number
- ◯ the smallest number
- ◯ the middle number
- ◯ the number that occurs most often

Place Value to 1,000

Beads, Buttons and Things

Use the Data

How many beads are in the jars?

What You Will Learn

In this chapter you will learn how to:

- Read, write, and show numbers to 1,000.

- Compare and order numbers to 1,000.

- Identify number patterns to 1,000.

- Find patterns to solve problems.

We Can Make Almost Anything

Story by Marina Ramos
Illustrated by Kristina Stephenson

MATH AT HOME

Dear Family,

In Chapter 12, I will learn about hundreds and numbers to 1,000. Here are vocabulary words and an activity we can do together.

Different Ways to Count

- Put about 200 beans in a bowl.

- Have your child find different ways to count the beans, such as by 5s, 10s or 100s.

- You may want to repeat the activity with other amounts of beans from 100 to 1,000.

Use

beans

beads

or buttons

Math Words

digits

4 2 5

the digits are
4 , 2, 5.

Place Value

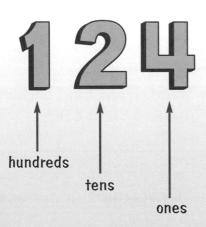

1 2 4

hundreds

tens

ones

Additional activities at
www.mhschool.com/math

McGraw-Hill School Division

Learn

Math Words

hundred

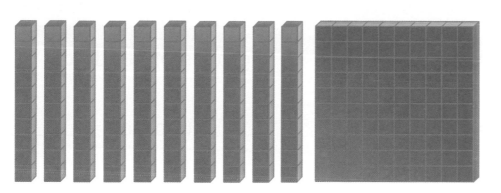

_10_ tens = _1_ hundred = _100_ in all

Try it

Use ▦ to make hundreds.
Write how many.

Workspace

1. 2 groups of hundreds

_2_ hundreds = _200_ in all

2. 3 groups of hundreds

_____ hundreds = _____ in all

3. 4 groups of hundreds

_____ hundreds = _____ in all

Sum it Up! How many hundreds are in 900? Explain.

_____

Math at Home: Your child used 100 squares to make groups of 100.
Activity: Ask your child to count by hundreds to 1,000.

Use to make hundreds.
Write how many.

4. 5 groups of hundreds

_5_ hundreds = _500_ in all

5. 6 groups of hundreds

____ hundreds = _____ in all

6. 7 groups of hundreds

____ hundreds = _____ in all

7. 8 groups of hundreds

____ hundreds = _____ in all

8. 9 groups of hundreds

____ hundreds = _____ in all

9. 10 groups of hundreds

____ hundreds = _____ in all

Problem Solving

Journal

Number Sense

10. Which group has more?
How do you know?

_____

_____

_____

Group A

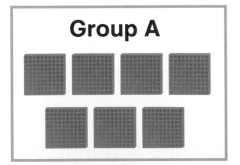

Group B

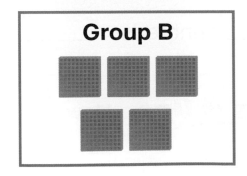

Hundreds, Tens, and Ones

Learn

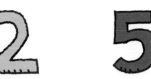
You can use hundreds, tens, and ones to show 325.

__3__ hundreds   __2__ tens   __5__ ones

hundreds	tens	ones
3	2	5

Try it   Write how many hundreds, tens, and ones.

1. 458

__4__ hundreds   __5__ tens   __8__ ones

hundreds	tens	ones
4	5	8

2. 142

_____ hundreds   _____ tens   _____ ones

hundreds	tens	ones

3. 636

_____ hundreds   _____ tens   _____ ones

hundreds	tens	ones

Sum it Up   How many hundreds, tens, and ones are in 572?

_____

Math at Home: Your child learned about hundreds, tens, and ones in 3-digit numbers.
Activity: Write a 3-digit number such as 291. Ask your child to tell how many hundreds, tens, and ones.

four hundred twenty-seven   **427**

Practice

Write how many hundreds, tens, and ones.

Remember to write a zero when there are no tens or ones.

4. 246

hundreds	tens	ones

_____ hundreds   _____ tens   _____ ones

5. 515

hundreds	tens	ones

_____ hundreds _____ tens   _____ ones

6. 630

hundreds	tens	ones

_____ hundreds   _____ tens   _____ ones

Write each number.

7. 3 hundreds 2 tens 6 ones _____

8. 5 hundreds 0 tens 8 ones _____

Critical Thinking

Journal

Draw a picture to solve.

Workspace

9. Lewis has 2 hundreds. Melanie has 3 hundreds. How many do they have in all?

_____ in all

MORE Hundreds, Tens, and Ones

Learn

You can learn the value of a digit by its place in a number.

hundreds	tens	ones
2	5	7

2 is the digit in the hundreds place.
5 is the digit in the tens place.
7 is the digit in the ones place.

Try it

Circle the value of each blue digit.

1. **347**   4 hundreds   (4 tens)   4 ones

2. **608**   6 hundreds   6 tens   6 ones

3. **923**   3 hundreds   3 tens   3 ones

4. **182**   8 hundreds   8 tens   8 ones

5. **256**   6 hundreds   6 tens   6 ones

6. **924**   9 hundreds   9 tens   9 ones

 **Sum it Up!**   How can you tell the value of a digit in a 3-digit number?

Math at Home: Your child learned the value of each digit in a 3-digit number.
Activity: Say a 3-digit number and have your child tell you the value of each digit.

7. 591   5 hundreds   5 tens   5 ones

8. 703   3 hundreds   3 tens   3 ones

9. 280   0 hundreds   0 tens   0 ones

10. 364   6 hundreds   6 tens   6 ones

11. 178   1 hundred   1 ten   1 one

12. 469   6 hundreds   6 tens   6 ones

13. 364   6 hundreds   6 tens   6 ones

14. 178   1 hundred   1 ten   1 one

15. 469   6 hundreds   6 tens   6 ones

16. 213   3 hundreds   3 tens   3 ones

17. 485   4 hundreds   4 tens   4 ones

18. 835   5 hundreds   5 tens   5 ones

Critical Thinking Journal

19. What is the greatest 3-digit number you can write using the digits 1, 7, and 5?

_____

Learn

You can count by hundreds.

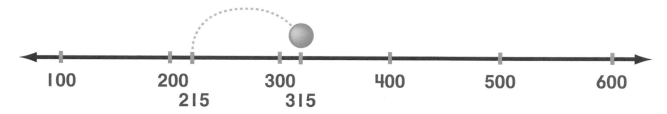

100   200   300   400   500   600
       215   315

One hundred more than 215 is 315.

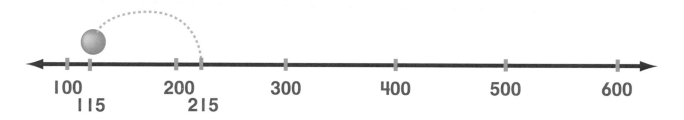

100   200   300   400   500   600
115   215

One hundred less than 215 is 115.

Try it   Write each number.

number	one hundred more
1. 573	673
2. 108	
3. 432	

number	one hundred less
573	473
108	
432	

 How do you find one hundred more than a number?

 Math at Home: Your child counted on or counted back by hundreds to find a number.
Activity: Choose a 3-digit number such as 427 or 165. Ask your child to
tell one hundred more than the number and one hundred less than the number.

Count on or back by hundreds.
Write each number.

Count on to find
one hundred more.
Count back to find
one hundred less.

4. One hundred more than 830 is __930__ .

5. One hundred less than 591 is _____ .

6. One hundred more than 152 is _____ .

7. One hundred less than 205 is _____ .

8. Two hundred more than 777 is _____ .

9. Three hundred less than 643 is _____ .

10. Four hundred more than 200 is _____ .

Problem Solving

Number Sense

11. Circle the box that has two hundred more toys than the bag.

Learn

You can write hundred with numbers and words.

hundreds	tens	ones
8	6	4

864

eight hundred sixty-four

You write this number.

You read this number.

Try it

Write each number. Write each word name.

1.

hundreds	tens	ones
3	1	7

317

three hundred seventeen

2.

hundreds	tens	ones
9	8	6

_____

_____

3.

hundreds	tens	ones
4	2	5

_____

_____

Sum it Up! How are the way you read a number and the way you write a number the same?

Math at Home: Your child read and wrote 3-digit numbers.
Activity: Write the digits 0 through 9 on index cards. Have your child choose three cards, arrange them to show a 3-digit number, and read the number aloud.

four hundred thirty-three

Write each number. Write each word.

4.

hundreds	tens	ones
3	1	8

318
three hundred eighteen

5.

hundreds	tens	ones
2	4	2

_____

_____

6.

hundreds	tens	ones
4	1	0

_____

_____

7.

hundreds	tens	ones
6	7	9

_____

_____

8.

hundreds	tens	ones
1	0	5

_____

_____

Problem Solving

Use Data

Use the picture. Solve.

9. How many buttons are there?
Write the number.
Write the word.

_____   _____

Learn

You can show a number in different ways.

expanded form

578 = 5 hundreds   7 tens   8 ones

500   +   70   +   8

Try it

Write each number in expanded form.

	Hundreds	Tens	Ones
1. 215	_200_	_10_	_5_
2. 690	___	___	___
3. 427	___	___	___
4. 766	___	___	___

How do you know if the 4 means 400 or 40 in the number 432?

Math at Home: Your child learned the value of each digit in a 3-digit number and wrote the number in expanded form.
Activity: Write some 3-digit numbers and have your child tell the value of each digit.

four hundred thirty-five   **435**

Write each number in expanded form.

	Hundreds	Tens	Ones
5. 455	400	50	5
6. 720	_____	_____	_____
7. 160	_____	_____	_____
8. 293	_____	_____	_____
9. 878	_____	_____	_____
10. 603	_____	_____	_____
11. 288	_____	_____	_____

Spiral Review and Test Prep

Choose the correct answer.

12. Which shows the number 965 in expanded form?

- ◯ 900 + 50 + 6
- ◯ 900 + 60 + 5
- ◯ 900 + 65 + 0
- ◯ 900 + 65 + 5

13. How many minutes are in one hour?

- ◯ 15 minutes
- ◯ 30 minutes
- ◯ 45 minutes
- ◯ 60 minutes

Name _____

Problem and Solutions

Reading Skill   You can find clues to help you solve problems.

Marcy is making a necklace. She chooses 5 green beads, 5 yellow beads, and 5 pink beads.

Solve.

1. How many pink beads are in the necklace? ____5____ pink beads

2. What is the pattern of Marcy's necklace?

_____

3. What color is the last bead that Marcy will put on the necklace?

_____

4. You have 15 beads. Color a pattern you can make.

Solve.

Jake is making
a bracelet.
He chooses 4 green
beads, 4 yellow beads,
and 8 orange beads.

5. How many orange beads are in the bracelet?
_____ orange beads

6. What is the pattern of Jake's bracelet?

_____

7. What color is the last bead that Jake will use?

_____

8. You have 16 beads. Color a pattern you can make.

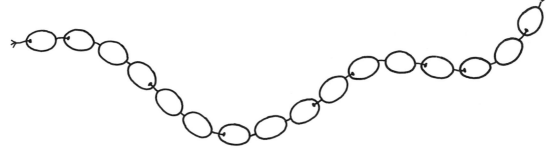

10. Write a problem about
making a pattern.

Check Your Progress A

Name _____

Write how many.

1. 3 groups of hundreds

_____ hundreds = _____ in all

2. 5 groups of hundreds

_____ hundreds = _____ in all

Write how many hundreds, tens, and ones.

3. 849

_____ hundreds   _____ tens   _____ ones

hundreds	tens	ones

4. 444

_____ hundreds   _____ tens   _____ ones

hundreds	tens	ones

Circle the value of the blue digit.

5. **3**28   2 hundreds   2 tens   2 ones

6. 9**0**1   9 hundreds   9 tens   9 ones

Write each number.

7. One hundred more than 298 is _____.

8. One hundred less than 861 is _____.

Write each number in expanded form.

9. 745   _____   +   _____   +   _____

10. 862   _____   +   _____   +   _____

Circle the two congruent figures.

1.

Write the fraction for the part that is shaded.

2.

 —— —— 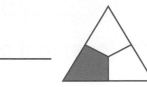 —— ——

Find each sum or difference.

3.

33	88	55	64	22¢	45¢
+ 24	− 27	+ 38	− 17	+ 32¢	− 15¢

TECHNOLOGY LINK

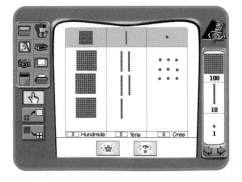

Place Value Models

- Choose place value.
- Choose a mat to show place value.
- Stamp out 3 hundreds, 5 tens, and 9 ones.
- What number is shown? _____

1. Choose place value. Show 2 hundreds, 4 tens, and 5 ones. What number is shown? _____

2. Stamp out other hundreds, tens, and ones. Tell the number shown.

For more practice use Math Traveler.™

Name _____

Find a Pattern

What do I know?

What do I need to find out?

Read ▶ Tina is making a necklace. She has put 9 beads on the string. What pattern will she use to complete the necklace?

Plan ▶ I can look at the necklace to see what the pattern is.

Solve ▶ The pattern is blue, blue, green. It repeats every 3 beads.

Look Back ▶ Does my answer make sense?

Continue a pattern. Draw the bead that could come next.

1.

_____

2.

_____

 How can you find a pattern in a string of beads?

Math at Home: Your child solved problems by finding and using a pattern.
Activity: Draw a simple red and green repeating pattern. Ask your child to draw what would most likely come next.

four hundred forty-one 441

McGraw-Hill School Division

Problem Solving · Strategy

Continue a pattern. Draw to show what would most likely come next.

3.

_____

4.

_____

5.

_____

6.

_____

7. Draw a picture.
   Make a bead bracelet.
   Make a pattern.
   Describe the pattern.

_____

_____

_____

Workspace

Compare Numbers

Learn

Compare numbers.

Look at the hundreds first. If the hundreds are the same, compare the tens.

You can compare numbers using <, >, and =.

231 is less than 253

 <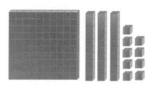

231 < 253

If the tens are the same, compare the ones.

139 is greater than 133

 >

139 > 133

326 is equal to 326

=

326 = 326

Try it   Compare. Write <, >, or =.

1. 374 197   425 425   621 628

2. 320 352   850  750   207 207

Sum it Up   How do you compare 3-digit numbers?

 **Math at Home:** Your child learned how to compare 3-digit numbers.
Activity: Ask your child to name three numbers that are greater than 286 and then three numbers that are less than 550.

Remember to compare the hundreds first.

3. 853 ⊜ 853   923 ◯ 927   567 ◯ 567

4. 689 ◯ 627   372 ◯ 374   450 ◯ 425

5. 281 ◯ 182   105 ◯ 105   789 ◯ 799

6. 601 ◯ 601   233 ◯ 230   955 ◯ 955

7. 723 ◯ 723   325 ◯ 300   252 ◯ 251

8. 533 ◯ 515   142 ◯ 180   697 ◯ 655

9. 487 ◯ 498   526 ◯ 526   190 ◯ 190

 Problem Solving

Logical Reasoning

10. I am greater than
3 hundreds 2 tens and
2 ones. I am less than
3 hundreds 2 tens and
4 ones. What number am I?

_____

11. I am greater than
8 hundreds 7 tens and
5 ones. I am less than
8 hundreds 7 tens and 7 ones.
What number am I?

_____

Learn

You can write numbers in order.

362·363·364·365·366·367·368·369·370·371·372

364 comes just before 365.

365 comes between 364 and 366.

366 comes just after 365.

Try it   Write the number that comes just before.

1. __250__ 251   ____ 685   ____ 450   ____ 702

Write the number that comes just after.

2. 354 ____   976 ____   555 ____   121 ____

Write the number that comes between.

3. 423 ____ 425   685 ____ 687   450 ____ 452

4. Which numbers come between 361 and 366? _____

Sum it Up   What is the number that comes just before 690?
What is the number that comes just after 690?

Math at Home: Your child ordered numbers.
Activity: Pick a number from 100 to 999. Have your child name
the numbers that come just before and just after the number.

four hundred forty-five **445**

Write the number that comes just before.

5. __871__ 872   ___ 103   ___ 724   ___ 465

6. ___ 517   ___ 110   ___ 333   ___ 199

7. ___ 303   ___ 200   ___ 421   ___ 219

Write the number that comes just after.

8. 350 ___   246 ___   429 ___   124 ___

9. 689 ___   766 ___   999 ___   532 ___

10. 152 ___   404 ___   300 ___   199 ___

Write the number that comes between.

11. 978 ___ 980   766 ___ 768   124 ___ 126

12. 322 ___ 324   769 ___ 771   800 ___ 802

Problem Solving

Number Sense

Circle your answer.

13. Which number is closest to 400?

    398   420   500

14. Which number is closest to 200?

    150   202   260

Learn

You can use number patterns to help you count.

350, 360, 370, 380

418, 518, 618, 718

I counted by tens.

I counted by hundreds.

Try it

Write the missing numbers.
Then circle the counting pattern.

Does the pattern show counting by hundreds, tens, or ones?

Numbers	Pattern: Count by
1. 140, 150, 160, <u>170</u>, 180, <u>190</u>	hundreds (tens) ones
2. 365, 465, ____, 665, ____, 865	hundreds tens ones
3. ____, 235, 236, 237, 238, ____	hundreds tens ones

 How can you tell if a number pattern is counting by hundreds?

 Math at Home: Your child described and extended number patterns.
Activity: Pick a number between 100 and 500. Have your child count by ones, tens, or hundreds.

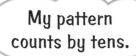

My pattern counts by tens.

428, 438, 448, 458, 468

Write the missing numbers in each pattern.
Then circle the counting pattern.

Numbers	Pattern: Count by
4. 920, 930, _940_, 950, 960, _____	hundreds (tens) ones
5. 432, _____, 632, _____, 832, 932	hundreds tens ones
6. 785, _____, 787, 788, 789, _____	hundreds tens ones
7. _____, 143, _____, 145, 146, 147	hundreds tens ones
8. _____, 510, 520, _____, 540, 550	hundreds tens ones
9. 299, _____, 499, 599, _____, 799	hundreds tens ones
10. 300, 400, _____, _____, 700, 800	hundreds tens ones

Problem Solving

11. The craft shop orders 200 boxes of beads each week. How many boxes of beads does the shop order in 5 weeks?

Number of Weeks	Boxes of Beads
1 Week	200 boxes
2 Weeks	400 boxes
3 Weeks	600 boxes
4 Weeks	_____ boxes
5 Weeks	_____ boxes

Name _____

Make a Bracelet

You want to make a bead bracelet
that shows a pattern.
Here are the beads you can use.

1. Select the beads you want
to use.
Make a list.
Draw a picture.

_____

_____

_____

_____

_____

Workspace

2. Draw some patterns you could make.
Then choose one.

Your Decision!

3. Draw a picture of your bead bracelet. Color it.

4. **What if** you made another bracelet?
Design some beads.
Then use your beads to
make a bracelet that
shows a pattern.

Name _____

How Well Does Sound Move Through String?

You Will Need

cups with holes
string
scissors
yardstick

Telephone wires carry sound from one place to another.

What to do

1. Work with a partner to make a string phone.

2. Measure a 30-foot string and cut it.

3. Put the string through the end of each cup and tie a knot.

4. Talk into the cup. Tell how well you can hear each other.

5. Repeat the activity with a 20-foot and a 10-foot string. How are the results alike and different?

STRING LENGTH	SOUND			COMMENTS
	great	OK	poor	
30 foot phone				
20 foot phone				
10 foot phone				

Did You KNOW?

Sound can move through matter such as air, metal and water.

What did you find out?

1. Which string length worked best? _____ feet

Explain why. _____

_____

2. How well do string phones work?

_____

3. What happens when string phones get longer?

_____

Journal

Want to do more?

What if you made the string 100 feet long?
How well do you think your string phone would work?

Name _____

Compare. Write >, <, or =.

1. 694 ◯ 694   472 ◯ 372   721 ◯ 728

2. 425 ◯ 415   998 ◯ 998   267 ◯ 274

Write the number that comes just before.

3. _____ 234   _____ 481   _____ 987

Write the number that comes just after.

4. 549 _____   699 _____   131 _____

Write the number that comes between.

5. 548 _____ 550   234 _____ 236   399 _____ 401

Write the missing numbers. Circle the counting pattern.

	Numbers	Pattern: Count by		
6.	520, 530, 540, _____, 560, _____	hundreds	tens	ones
7.	234, 334, _____, 534, _____, 734	hundreds	tens	ones
8.	_____, 331, 332, 333, 334, _____	hundreds	tens	ones
9.	680, 690, _____, 710, 720, _____	hundreds	tens	ones
10.	120, 220, _____, 420, _____, 620	hundreds	tens	ones

Name _____

What Am I?

Color these squares to make a picture.

 Numbers between:
312 and 315
372 and 375

 Numbers between:
312 and 315
372 and 375

 Numbers between:
330 and 337
350 and 357

 8 numbers
before 348

 Number between:
383 and 385
5 numbers after 394

300	301	302	303	304	305	306	307	308	309
310	311	312	313	314	315	316	317	318	319
320	321	322	323	324	325	326	327	328	329
330	331	332	333	334	335	336	337	338	339
340	341	342	343	344	345	346	347	348	349
350	351	352	353	354	355	356	357	358	359
360	361	362	363	364	365	366	367	368	369
370	371	372	373	374	375	376	377	378	379
380	381	382	383	384	385	386	387	388	389
390	391	392	393	394	395	396	397	398	399

I fly in the air with a string for a tail. _____

Chapter Review

Name _____

Language and Math

Complete.
Use words from the list.

Read these words.

Math Words

greater than
less than
digit
hundreds
tens

1. 362 is _____ 363.

2. 975 is _____ 246.

3. In the number 345, the value of the _____ is 4 tens.

4. In the number 159, the 1 is in the _____ place.

Concepts and Skills

Write how many hundreds, tens, and ones.

5. 458

    _____ hundreds _____ tens   _____ ones

hundreds	tens	ones

6. 142

    _____ hundreds _____ tens   _____ ones

hundreds	tens	ones

Circle the value of the blue digit.

7. **5**47   4 hundreds   4 tens   4 ones

8. **7**08   7 hundreds   7 tens   7 ones

9. 2**4**5   5 hundreds   5 tens   5 ones

Compare. Write <, >, or =.

10. 769 ◯ 770   545 ◯ 445   356 ◯ 356

Write the number that comes just before.

11. _____ 436   _____ 880   _____ 122

Write the number that comes just after.

12. 649 _____   399 _____   925 _____

Write the number that comes between.

13. 321 _____ 323   985 _____ 987   569 _____ 571

Write the missing numbers.
Then circle the counting pattern.

Number	Pattern: Count by		
14. 921, 922, _____, 924, 925, _____	hundreds	tens	ones
15. 220, _____, 420, _____, 620, 720	hundreds	tens	ones
16. 535, _____, 555, 565, 575, _____	hundreds	tens	ones

Problem Solving

17. Continue the pattern. Draw to show what would

most likely come next. _____

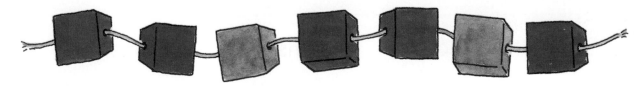

I'm thinking of a number.

Write a 3-digit number.	
How many hundreds are in your number?	
How many tens are in your number?	
How many ones are in your number?	
Write your number in expanded form.	
Name a number that is greater than your number.	
Name a number that is less than your number.	
Name a number that is equal to your number.	
What number comes just after your number?	
What number comes just before your number?	
What number is one hundred more than your number?	
What number is one hundred less than your number?	

You can arrange the same three digits
to make different numbers.

hundreds	tens	ones	number
4	6	8	468
6	4	8	648

Arrange the cards in different ways.
Make as many numbers as possible.

hundreds	tens	ones	number

Name _____

Write how many hundreds, tens, and ones.

1. 400

hundreds	tens	ones
4	0	0

_____ hundreds _____ tens _____ ones

Circle the value of the blue digit.

2. 378   7 hundreds   7 tens   7 ones

Write each number. Write each word.

3. One hundred more than 824 is _____.

4. One hundred less than 824 is _____.

Compare. Write >, <, or =.

5. 545 ◯ 645   234 ◯ 234   333 ◯ 233

Write the missing numbers in each pattern.
Circle the pattern you used to count.

6. 225, 325, 425, _____, 625, _____	hundreds tens ones	
7. 434, _____, 454, 464, _____, 484	hundreds tens ones	

8. Continue the pattern. Write what could come next.

_____

McGraw-Hill School Division

Name _____

Toss 3 cubes onto the number mat.
Use the three digits to make the
greatest number possible.
Record the number. Repeat.
Use <, >, or = to compare the numbers.

You Will Need

3 cubes

is less than
< ,

is greater than
>

is equal to
=

	Number	<, >, or =	Number
1.		◯	
2.		◯	
3.		◯	
4.		◯	
5.		◯	
6.		◯	
7.		◯	
8.		◯	

 Portfolio You may want to put this page in your portfolio.

Name _____

Choose the correct answer.

Mathematical Reasoning

1. What is the next number in this skip counting pattern?

20, 30, 40, 50, _____

Explain how you know.

_____

_____

2. Which tool would you use to measure the length of your bike?

- ○ ruler
- ○ thermometer
- ○ scale
- ○ cup

3. Miles had $3.40. He finds one coin. Now he has $3.65. Which coin did he find?

- ○ penny
- ○ nickel
- ○ dime
- ○ quarter

Number Sense

4. Which number shows five hundred seventy-three?

- ○ 503
- ○ 537
- ○ 573
- ○ 773

5. Which number is one hundred more than 829?

- ○ 729
- ○ 730
- ○ 839
- ○ 929

6. There are 360 blue beads. There are 150 yellow beads. How many more blue beads than yellow beads are there?

- ○ 510
- ○ 310
- ○ 210
- ○ 200

Algebra and Functions

Use the graph to answer questions 7 and 8.

Carrie's Collection

keys	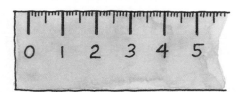
buttons	
beads	
coins	

7. How many keys and beads does Carrie have?

○ 11
○ 14
○ 8
○ 9

8. How many more buttons than coins are there?

○ 4
○ 1
○ 9
○ 3

9. Sue has 27 stickers. Nat gives her 18 more stickers. Which shows how many stickers Sue has now?

○ 27 + 18
○ 27 − 18
○ 45 + 27
○ 45 − 18

Measurement and Geometry

10. How many centimeters long is the string of beads?

○ 5 centimeters
○ 1 centimeter
○ 4 centimeters
○ 3 centimeters

11. What is the time?

○ 4:15
○ 3:30
○ 4:45
○ 3:45

12. Which figure is a pyramid?

theme

Community Helpers

GOAL $999
RAISED SO FAR
$565

Use the Data

How much more money needs to be raised to meet the goal?

What You Will Learn

In this chapter you will learn how to:

- Add and subtract 3-digit numbers.

- Add and subtract money amounts.

- Make a graph to solve problems.

Community Helper Week

Story by Vivian Abraite
Illustrated by Mike Danner

four hundred sixty-three **463**

Math At Home

McGraw-Hill School Division

Dear Family,

In Chapter 13, I will learn how to add and subtract 3-digit numbers. Here are vocabulary words and an activity that we can do together.

Take Tens!

Math Words

addend

$$
\begin{array}{r}
231 \\
+\ 318 \\
\hline
549
\end{array}
$$

addends

hundreds

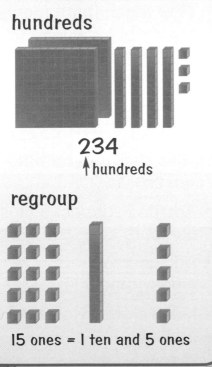

234

↑ hundreds

regroup

15 ones = 1 ten and 5 ones

- Fold a piece of paper into 3 columns. Label the columns "hundreds," "tens," and "ones."

hundreds	tens	ones

- Write several numbers between 100 and 999 on separate cards.

- Ask your child to choose cards and write the numbers as hundreds, tens, and ones.

use

pencil and paper

index cards

Additional activities at www.mhschool.com/math

Learn

You can count on to add hundreds.

Start on 500.

Count on 3 hundreds.

$$500 + 300 = \underline{800}$$

Start at 500. Count on 3 hundreds.
600, 700, 800

Try it   Count on by hundreds to add.

1. $100 + 500 = \underline{600}$ $200 + 300 = \underline{500}$

2. $500 + 200 = \underline{}$ $400 + 200 = \underline{}$

3.

100	300	200	100	400	100
+ 300	+ 400	+ 300	+ 600	+ 100	+ 800

4.

200	700	300	200	500	400
+ 100	+ 200	+ 500	+ 400	+ 100	+ 300

 **Sum it Up!**   How can you find $400 + 300$?

 Math at Home: Your child added hundreds by counting on.
Activity: Have your child show you how to count on to add $600 + 200$.

5.

300	400	200	400	600	500
+ 100	+ 300	+ 400	+ 400	+ 300	+ 400

6.

500	100	300	200	400	600
+ 300	+ 500	+ 300	+ 500	+ 100	+ 200

7.

600	500	100	300	200	400
+ 100	+ 200	+ 200	+ 600	+ 700	+ 500

8.

300	300	500	100	300	200
+ 200	+ 400	+ 100	+ 400	+ 500	+ 200

Problem Solving

Mental Math

9. The school collects cans to recycle. The second graders collect 300 cans. The third graders collect 500 cans. How many cans do they collect altogether?

_____ cans

10. The second grade read 400 books. The third grade read 500 books. How many books did they read altogether?

_____ books

Add.

1. 9 + 7 = ——

2. 8 + 6 = ——

3. 5 + 8 = ——

4. 9 + 8 = ——

5. 6 + 9 = ——

6. 10 + 10 = ——

7. 6 + 8 = ——

8. 8 + 7 = ——

9. 7 + 9 = ——

10. 9 + 9 = ——

11. 7 + 8 = ——

12. 10 + 9 = ——

13. 6 + 7 = ——

14. 9 + 6 = ——

15. 7 + 0 = ——

16. 5 + 9 = ——

17. 9 + 5 = ——

18. 7 + 6 = ——

19. 7 + 7 = ——

20. 8 + 5 = ——

21. 8 + 8 = ——

22. 8 + 9 = ——

23. 4 + 9 = ——

24. 9 + 4 = ——

Math at Home: Your child practiced addition facts.
Activity: Cover the answers with a paper strip. Time your child as he or she writes the answers again. You can repeat daily to help your child recall the facts quickly.

Facts Practice: Addition and Subtraction

Add or subtract.

1.

$$\begin{array}{r} 6 \\ +\ 9 \\ \hline \end{array} \qquad \begin{array}{r} 12 \\ -\ 8 \\ \hline \end{array} \qquad \begin{array}{r} 8 \\ -\ 8 \\ \hline \end{array} \qquad \begin{array}{r} 8 \\ +\ 5 \\ \hline \end{array} \qquad \begin{array}{r} 12 \\ -\ 9 \\ \hline \end{array} \qquad \begin{array}{r} 6 \\ +\ 8 \\ \hline \end{array} \qquad \begin{array}{r} 10 \\ +\ 6 \\ \hline \end{array}$$

2.

$$\begin{array}{r} 9 \\ -\ 5 \\ \hline \end{array} \qquad \begin{array}{r} 8 \\ +\ 6 \\ \hline \end{array} \qquad \begin{array}{r} 11 \\ -\ 9 \\ \hline \end{array} \qquad \begin{array}{r} 8 \\ -\ 4 \\ \hline \end{array} \qquad \begin{array}{r} 8 \\ +\ 8 \\ \hline \end{array} \qquad \begin{array}{r} 11 \\ -\ 7 \\ \hline \end{array} \qquad \begin{array}{r} 7 \\ +\ 6 \\ \hline \end{array}$$

3.

$$\begin{array}{r} 5 \\ +\ 9 \\ \hline \end{array} \qquad \begin{array}{r} 10 \\ -\ 7 \\ \hline \end{array} \qquad \begin{array}{r} 11 \\ -\ 8 \\ \hline \end{array} \qquad \begin{array}{r} 7 \\ +\ 8 \\ \hline \end{array} \qquad \begin{array}{r} 9 \\ -\ 9 \\ \hline \end{array} \qquad \begin{array}{r} 7 \\ +\ 9 \\ \hline \end{array} \qquad \begin{array}{r} 15 \\ -\ 8 \\ \hline \end{array}$$

4.

$$\begin{array}{r} 4 \\ +\ 9 \\ \hline \end{array} \qquad \begin{array}{r} 7 \\ -\ 4 \\ \hline \end{array} \qquad \begin{array}{r} 12 \\ -\ 6 \\ \hline \end{array} \qquad \begin{array}{r} 11 \\ -\ 5 \\ \hline \end{array} \qquad \begin{array}{r} 8 \\ +\ 9 \\ \hline \end{array} \qquad \begin{array}{r} 10 \\ -\ 2 \\ \hline \end{array} \qquad \begin{array}{r} 16 \\ -\ 8 \\ \hline \end{array}$$

5.

$$\begin{array}{r} 9 \\ +\ 9 \\ \hline \end{array} \qquad \begin{array}{r} 8 \\ -\ 3 \\ \hline \end{array} \qquad \begin{array}{r} 17 \\ -\ 9 \\ \hline \end{array} \qquad \begin{array}{r} 7 \\ +\ 7 \\ \hline \end{array} \qquad \begin{array}{r} 10 \\ -\ 8 \\ \hline \end{array} \qquad \begin{array}{r} 3 \\ +\ 8 \\ \hline \end{array} \qquad \begin{array}{r} 9 \\ +\ 6 \\ \hline \end{array}$$

6.

$$\begin{array}{r} 10 \\ -\ 6 \\ \hline \end{array} \qquad \begin{array}{r} 6 \\ +\ 7 \\ \hline \end{array} \qquad \begin{array}{r} 16 \\ -\ 9 \\ \hline \end{array} \qquad \begin{array}{r} 8 \\ +\ 7 \\ \hline \end{array} \qquad \begin{array}{r} 4 \\ +\ 8 \\ \hline \end{array} \qquad \begin{array}{r} 15 \\ -\ 6 \\ \hline \end{array} \qquad \begin{array}{r} 12 \\ -\ 5 \\ \hline \end{array}$$

7.

$$\begin{array}{r} 9 \\ +\ 4 \\ \hline \end{array} \qquad \begin{array}{r} 14 \\ -\ 7 \\ \hline \end{array} \qquad \begin{array}{r} 11 \\ -\ 6 \\ \hline \end{array} \qquad \begin{array}{r} 7 \\ +\ 5 \\ \hline \end{array} \qquad \begin{array}{r} 9 \\ +\ 7 \\ \hline \end{array} \qquad \begin{array}{r} 18 \\ -\ 9 \\ \hline \end{array} \qquad \begin{array}{r} 14 \\ -\ 8 \\ \hline \end{array}$$

Name_____

Learn

You can use
addition facts
to help.

324 second graders pulled weeds in the park.
261 third graders planted flowers.
How many children worked in the park?

Add to find how many in all.

1 Add the ones.

hundreds	tens	ones
3	2	4
+ 2	6	1
		5

2 Add the tens.

hundreds	tens	ones
3	2	4
+ 2	6	1
	8	5

3 Add the hundreds.

hundreds	tens	ones
3	2	4
+ 2	6	1
5	8	5

585 children in all

Try it   Find each sum.

1.

hundreds	tens	ones
1	2	4
+ 4	5	3
5	7	7

hundreds	tens	ones
3	5	1
+	2	7

hundreds	tens	ones
2	4	3
+		5

 **Sum it Up!**   What addition facts can help you add 362 + 314?

 Math at Home: Your child used basic facts to add hundreds, tens, and ones.
Activity: Have your child add 263 and 426.

four hundred sixty-nine **469**

Add the ones first.

2.

hundreds	tens	ones
6	2	5
+ 2	4	3
8	6	8

hundreds	tens	ones
2	8	4
+	1	5

hundreds	tens	ones
1	6	2
+		7

3.

406	216	712	238	413	215
+ 381	+ 32	+ 7	+ 61	+ 265	+ 462

4.

324	652	431	222	194	523
+ 260	+ 26	+ 253	+ 5	+ 303	+ 74

5.

147	244	631	324	427	503
+ 32	+ 504	+ 7	+ 302	+ 21	+ 274

Problem Solving

Number Sense

6. What is the same about 523 and 325?
What is different?

Learn

Add 356 + 281.

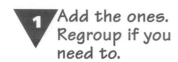 **1** Add the ones. Regroup if you need to.

hundreds	tens	ones
☐	☐	☐
3	5	6
+ 2	8	1
		7

7 ones

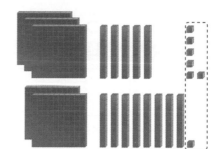

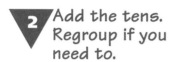 **2** Add the tens. Regroup if you need to.

hundreds	tens	ones
☐	☐	☐
3	5	6
+ 2	8	1
	3	7

13 tens =
1 hundred 3 tens

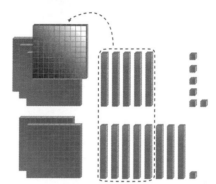

3 Add the hundreds.

hundreds	tens	ones
1	☐	☐
3	5	6
+ 2	8	1
6	3	7

6 hundreds

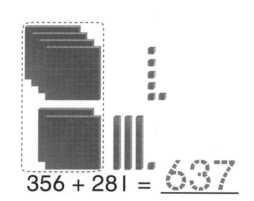

356 + 281 = _637_

Try it

Find each sum. Use hundreds, tens, and ones models.

1.

hundreds	tens	ones
1	☐	☐
4	5	2
+ 3	8	5
8	3	7

hundreds	tens	ones
☐	☐	☐
5	2	9
+ 2	3	7

hundreds	tens	ones
☐	☐	☐
4	6	8
+ 3	2	2

 **Sum it Up** How do you know when to regroup?

 Math at Home: Your child added numbers, deciding each time if regrouping was needed.
Activity: Have your child show you how to add 537 + 291 and explain the regrouping to you.

Practice Add.

> Remember, there are
> 10 tens in 1 hundred.

2.

hundreds	tens	ones
☐ 1	☐	
5	3	5
+ 2	8	4
8	1	9

hundreds	tens	ones
☐	☐	
3	7	6
+ 4	3	2

hundreds	tens	ones
☐	☐	
2	1	5
+ 1	3	5

3.

hundreds	tens	ones
☐	☐	
3	8	2
+ 2	6	5

hundreds	tens	ones
☐	☐	
4	2	6
+ 3	4	9

hundreds	tens	ones
☐	☐	
2	7	4
+ 3	6	2

4.

hundreds	tens	ones
☐	☐	
2	6	5
+ 1	2	7

hundreds	tens	ones
☐	☐	
5	3	7
+ 2	4	6

hundreds	tens	ones
☐	☐	
3	4	6
+ 1	9	2

Problem Solving

Draw a Picture

5. Use models to show the sum of 120 + 205.

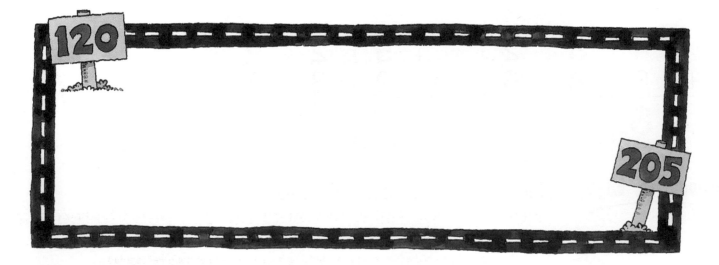

Name _____

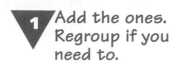 **Learn**

Sometimes you have to regroup ones and tens.

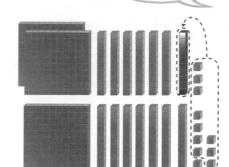

Add 263 + 178.

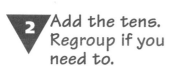 **1** Add the ones. Regroup if you need to.

hundreds	tens	ones
☐	☑ 1	☐
2	6	3
+ 1	7	8
		1

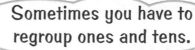

2 Add the tens. Regroup if you need to.

hundreds	tens	ones
☑ 1	☑ 1	☐
2	6	3
+ 1	7	8
	4	1

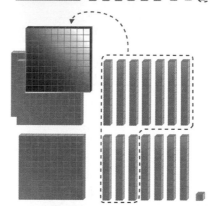

3 Add the hundreds.

hundreds	tens	ones
☑ 1	☑ 1	☐
2	6	3
+ 1	7	8
4	4	1

263 + 178 = __441__

Try it Add.

1.

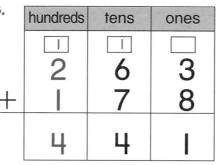

hundreds	tens	ones
☐	☐	☐
5	1	8
+ 2	8	9
8	0	7

hundreds	tens	ones
☐	☐	☐
1	8	9
+ 3	7	8

hundreds	tens	ones
☐	☐	☐
4	7	6
+ 1	4	4

 **Sum it Up** How can you tell whether or not you need to regroup?

 Math at Home: Your child learned how to add 3-digit numbers with and without regrouping.
Activity: Have your child show you an addition problem that needs regrouping and a problem that does not need regrouping.

four hundred seventy-three **473**

Add. Regroup if you need to.

2.

hundreds	tens	ones
☐	☐	☐
4	5	7
+ 1	8	5
6	4	2

hundreds	tens	ones
☐	☐	☐
5	6	9
+ 1	7	4

hundreds	tens	ones
☐	☐	☐
3	6	8
+ 2	9	4

3.

264	425	379	264	108	216
+ 253	+ 342	+ 148	+ 288	+ 419	+ 405

4.

332	115	163	423	662	449
+ 146	+ 248	+ 249	+ 217	+ 109	+ 138

5.

223	216	139	402	329	175
+ 415	+ 451	+ 240	+ 163	+ 261	+ 230

Problem Solving

6. Lincoln School had a paper drive. How many pounds of paper did Grade 2 and Grade 3 collect together?

_____ pounds

Newspapers Collected	
Grade	Pounds
Grade 1	245
Grade 2	326
Grade 3	314

Estimate Sums

Learn

Rewrite. Use nearest hundreds.

Clarissa's class collected 375 cans in May.
They collected 220 bottles in June.
How many cans and bottles did the
class collect?

$$375 + 220 = 595$$

The class collected 595 cans and bottles.
You can add nearest hundreds to see if your answer is reasonable.

| 200 | 220 | 250 | | 300 | | 350 | 375 | 500 | | 450 | | 500 | | 550 | | 600 |

$$
\begin{array}{r}
375 \\
+ 220 \\
\hline
595
\end{array}
\quad
\begin{array}{l}
\text{nearest hundred} \\
\text{nearest hundred}
\end{array}
\quad
\begin{array}{r}
400 \\
+ 200 \\
\hline
600
\end{array}
$$

595 is close to 600.
The answer is reasonable.

Try it

Add. Estimate to see if your answer is reasonable.

1.

$$
\begin{array}{r}
427 \\
+ 386 \\
\hline
813
\end{array}
\qquad
\begin{array}{r}
400 \\
+ 400 \\
\hline
800
\end{array}
\qquad
\begin{array}{r}
529 \\
+ 263 \\
\hline
\end{array}
\qquad
\begin{array}{r}
 \\
+ \\
\hline
\end{array}
$$

2.

$$
\begin{array}{r}
368 \\
+ 218 \\
\hline
\end{array}
\qquad
\begin{array}{r}
 \\
+ \\
\hline
\end{array}
\qquad
\begin{array}{r}
212 \\
+ 489 \\
\hline
\end{array}
\qquad
\begin{array}{r}
 \\
+ \\
\hline
\end{array}
$$

Sum it Up

How can you estimate to check your addition?

Math at Home: Your child estimated sums to check addition by finding the nearest hundred.
Activity: Have your child add 478 + 329 and then estimate to see if
the answer is reasonable.

four hundred seventy-five **475**

Add. Estimate to see if your answer is reasonable.

3.
$$\begin{array}{r} 673 \\ +\,184 \\ \hline 857 \end{array}$$
$$\begin{array}{r} 700 \\ +200 \\ \hline 900 \end{array}$$
$$\begin{array}{r} 269 \\ +\,307 \\ \hline \end{array}$$
$$\begin{array}{r} \\ + \\ \hline \end{array}$$

4.
$$\begin{array}{r} 283 \\ +\,485 \\ \hline \end{array}$$
$$\begin{array}{r} \\ + \\ \hline \end{array}$$
$$\begin{array}{r} 319 \\ +\,126 \\ \hline \end{array}$$
$$\begin{array}{r} \\ + \\ \hline \end{array}$$

5.
$$\begin{array}{r} 329 \\ +\,592 \\ \hline \end{array}$$
$$\begin{array}{r} \\ + \\ \hline \end{array}$$
$$\begin{array}{r} 423 \\ +\,385 \\ \hline \end{array}$$
$$\begin{array}{r} \\ + \\ \hline \end{array}$$

6.
$$\begin{array}{r} 202 \\ +\,465 \\ \hline \end{array}$$
$$\begin{array}{r} \\ + \\ \hline \end{array}$$
$$\begin{array}{r} 416 \\ +\,328 \\ \hline \end{array}$$
$$\begin{array}{r} \\ + \\ \hline \end{array}$$

Spiral Review and Test Prep

Choose the correct answer.

7.
$$\begin{array}{r} 335 \\ +\,265 \\ \hline \end{array}$$

- ◯ 200
- ◯ 300
- ◯ 500
- ◯ 600

8. Which fraction names the part that is shaded?

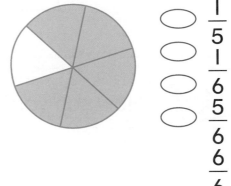

- ◯ $\dfrac{1}{5}$
- ◯ $\dfrac{1}{6}$
- ◯ $\dfrac{5}{6}$
- ◯ $\dfrac{6}{6}$

Name _____

Main Idea and Details

Reading You can find the main idea and
Skill details to solve problems.

The library asked schools
for book donations.
Lenora's class donated
217 picture books. Billy's
class donated 198
adventure books.

Solve.

1. What did the children do? _____

2. What do you want to find out? _____

3. What details will help you find out? _____

4. How many books do the children donate? _____ books

5. Write a number sentence to show your thinking.

_____

Math at Home: Your child used main ideas and details to solve a problem.
Activity: Tell your child a simple number story. Then have your child
tell you the main idea and details.

four hundred seventy-seven **477**

Solve.

Sammy and Sally collect newspapers to recycle. Sammy collects 132 newspapers. Sally collects 119 newspapers.

6. What is the main idea? _____

7. What do you want to find out? _____

8. What details will help you find out? _____

_____

9. How many newspapers do they collect? _____ newspapers

10. Write a number sentence to show your thinking.

_____

11. Use the story. Write a problem.

_____

Check Your Progress A

Name _____

Add.

1.
$$\begin{array}{r} 100 \\ +\ 200 \\ \hline \end{array}$$
$$\begin{array}{r} 200 \\ +\ 400 \\ \hline \end{array}$$
$$\begin{array}{r} 300 \\ +\ 400 \\ \hline \end{array}$$
$$\begin{array}{r} 200 \\ +\ 600 \\ \hline \end{array}$$
$$\begin{array}{r} 300 \\ +\ 100 \\ \hline \end{array}$$

2.
$$\begin{array}{r} 243 \\ +\ 124 \\ \hline \end{array}$$
$$\begin{array}{r} 305 \\ +\ 232 \\ \hline \end{array}$$
$$\begin{array}{r} 422 \\ +\ 247 \\ \hline \end{array}$$
$$\begin{array}{r} 530 \\ +\ 247 \\ \hline \end{array}$$
$$\begin{array}{r} 342 \\ +\ 134 \\ \hline \end{array}$$

3.
$$\begin{array}{r} 251 \\ +\ 738 \\ \hline \end{array}$$
$$\begin{array}{r} 475 \\ +\ 318 \\ \hline \end{array}$$
$$\begin{array}{r} 629 \\ +\ 147 \\ \hline \end{array}$$
$$\begin{array}{r} 572 \\ +\ 294 \\ \hline \end{array}$$
$$\begin{array}{r} 336 \\ +\ 285 \\ \hline \end{array}$$

4.
$$\begin{array}{r} 294 \\ +\ 375 \\ \hline \end{array}$$
$$\begin{array}{r} 427 \\ +\ 328 \\ \hline \end{array}$$
$$\begin{array}{r} 587 \\ +\ 249 \\ \hline \end{array}$$
$$\begin{array}{r} 231 \\ +\ 463 \\ \hline \end{array}$$
$$\begin{array}{r} 473 \\ +\ 184 \\ \hline \end{array}$$

5.
$$\begin{array}{r} 365 \\ +\ 299 \\ \hline \end{array}$$
$$\begin{array}{r} 216 \\ +\ 487 \\ \hline \end{array}$$
$$\begin{array}{r} 427 \\ +\ 492 \\ \hline \end{array}$$
$$\begin{array}{r} 391 \\ +\ 133 \\ \hline \end{array}$$
$$\begin{array}{r} 202 \\ +\ 716 \\ \hline \end{array}$$

Add or subtract.

1.

58	63¢	47	26	35	74
+ 19	− 47¢	+ 28	− 18	+ 49	− 31

Write each number as tens and ones and then in expanded form.

2. 85 = _____ tens _____ ones = _____ + _____

Write the fraction for the shaded part.

3. 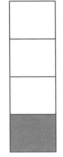 _____ 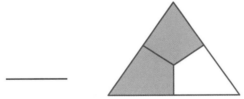 _____ 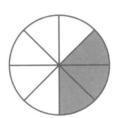 _____

 **TECHNOLOGY LINK**

Use the Internet

- Go to www.mhschool.com/math.
- Find the site about cats and dogs.
- Click on the link.
- What is the annual feeding cost for a cat? _____
- What is the annual feeding cost for two cats? _____

 Why is the Internet a good place to find information?

For more practice use Math Traveler.™

Name _____

Make a Graph

What do I know?

What do I need to find out?

Read
Cindy handed out 12 sports drinks, 10 bottles of water, and 6 cups of juice to walkers at the Book Walk.
Harry handed out 6 sport drinks, 12 waters, and 4 juices.

Which drink was handed out most often?

Plan
I can make a bar graph to solve.

Solve

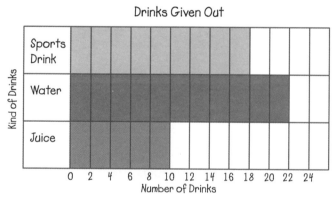

Drinks Given Out

Kind of Drinks: Sports Drink, Water, Juice

Number of Drinks: 0 2 4 6 8 10 12 14 16 18 20 22 24

The graph shows that Cindy and Harry handed out 18 sports drinks, 22 waters, and 10 juices.

Water was handed out most often.

Look Back
Does your answer make sense? Explain.

How does making a graph help you solve the problem?

Math at Home: Your child solved problems by making graphs.
Activity: Have your child explain the Book Walk problem.

Solve. Complete the graph.

1. In the Book Walk, Jon walked the first lap in 15 minutes and the next lap in 20 minutes. Betty walked the first lap in 10 minutes and the next lap in 25 minutes. Kathy walked the first lap in 10 minutes and the next lap in 10 minutes. How long did each walk altogether?

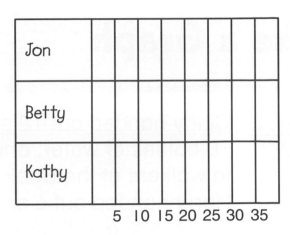

_____

_____

2. Yu Lee brought 3 pink flowers and 2 white flowers for the walkers. Jennifer brought 2 blue flowers and 5 red flowers. Rebecca brought 3 yellow flowers and 3 orange flowers. How many flowers did each girl bring?

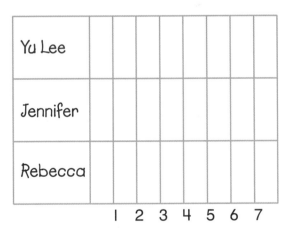

_____

_____

3. Write a problem for this graph.

_____

Book Walk

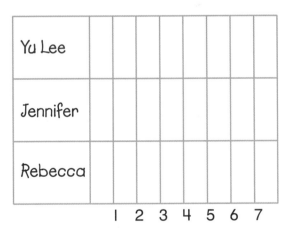

Learn

You can count back to subtract.

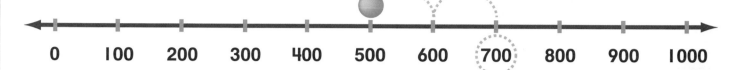

0   100   200   300   400   500   600   700   800   900   1000

$700 - 200 = $ _500_

Start at 700.
Count back
2 hundreds.
600, 500

Try it   Count back by hundreds to subtract.

1. $400 - 100 = $ _300_ $500 - 300 = $ _200_

2. $500 - 200 = $ ___ $700 - 100 = $ ___

3.
$$\begin{array}{r} 800 \\ -\ 300 \\ \hline \end{array} \quad \begin{array}{r} 600 \\ -\ 100 \\ \hline \end{array} \quad \begin{array}{r} 500 \\ -\ 300 \\ \hline \end{array} \quad \begin{array}{r} 900 \\ -\ 200 \\ \hline \end{array} \quad \begin{array}{r} 700 \\ -\ 300 \\ \hline \end{array} \quad \begin{array}{r} 300 \\ -\ 100 \\ \hline \end{array}$$

4.
$$\begin{array}{r} 600 \\ -\ 300 \\ \hline \end{array} \quad \begin{array}{r} 400 \\ -\ 200 \\ \hline \end{array} \quad \begin{array}{r} 900 \\ -\ 300 \\ \hline \end{array} \quad \begin{array}{r} 500 \\ -\ 100 \\ \hline \end{array} \quad \begin{array}{r} 200 \\ -\ 100 \\ \hline \end{array} \quad \begin{array}{r} 300 \\ -\ 200 \\ \hline \end{array}$$

 How can you subtract $800 - 200$?

Math at Home: Your child subtracted hundreds by counting back.
Activity: Have your child show you how to count back to subtract
$500 - 200$.

four hundred eighty-three   **483**

Practice Subtract.

Remember to count back by hundreds.

5. $500 - 100 = \underline{400}$ $600 - 200 = \underline{400}$

6. $700 - 200 = \underline{}$ $500 - 300 = \underline{}$

7. $900 - 300 = \underline{}$ $800 - 200 = \underline{}$

8.
$$\begin{array}{r} 300 \\ -\ 200 \\ \hline \end{array} \qquad \begin{array}{r} 500 \\ -\ 300 \\ \hline \end{array} \qquad \begin{array}{r} 400 \\ -\ 200 \\ \hline \end{array} \qquad \begin{array}{r} 800 \\ -\ 200 \\ \hline \end{array} \qquad \begin{array}{r} 700 \\ -\ 400 \\ \hline \end{array} \qquad \begin{array}{r} 600 \\ -\ 200 \\ \hline \end{array}$$

9.
$$\begin{array}{r} 800 \\ -\ 400 \\ \hline \end{array} \qquad \begin{array}{r} 700 \\ -\ 300 \\ \hline \end{array} \qquad \begin{array}{r} 900 \\ -\ 400 \\ \hline \end{array} \qquad \begin{array}{r} 600 \\ -\ 100 \\ \hline \end{array} \qquad \begin{array}{r} 500 \\ -\ 100 \\ \hline \end{array} \qquad \begin{array}{r} 600 \\ -\ 300 \\ \hline \end{array}$$

Problem Solving

Algebra & functions   Find the next number in each pattern.

10. | 400 500 600 700 ____ | 235 335 435 535 ____ |

11. | 800 700 600 500 ____ | 926 826 726 626 ____ |

484 four hundred eighty-four

Name_____

Subtract.

1. $17 - 9 =$ _____

2. $13 - 7 =$ _____

3. $14 - 4 =$ _____

4. $13 - 5 =$ _____

5. $17 - 8 =$ _____

6. $15 - 5 =$ _____

7. $13 - 4 =$ _____

8. $14 - 9 =$ _____

9. $12 - 3 =$ _____

10. $13 - 3 =$ _____

11. $13 - 8 =$ _____

12. $15 - 9 =$ _____

13. $16 - 9 =$ _____

14. $15 - 8 =$ _____

15. $18 - 9 =$ _____

16. $14 - 7 =$ _____

17. $16 - 7 =$ _____

18. $15 - 7 =$ _____

19. $13 - 9 =$ _____

20. $14 - 6 =$ _____

21. $13 - 6 =$ _____

22. $15 - 6 =$ _____

23. $16 - 8 =$ _____

24. $14 - 8 =$ _____

McGraw-Hill School Division

Math at Home: Your child practiced subtraction facts.
Activity: Cover the answers with a paper strip. Time your child as he
or she writes the answers again. You can repeat daily to help your
child recall the facts quickly.

four hundred eighty-five   485

Facts Practice: Addition and Subtraction

Add or subtract.

1.

$$\begin{array}{r} 6 \\ + 7 \\ \hline \end{array}\qquad \begin{array}{r} 9 \\ - 9 \\ \hline \end{array}\qquad \begin{array}{r} 12 \\ - 9 \\ \hline \end{array}\qquad \begin{array}{r} 9 \\ + 6 \\ \hline \end{array}\qquad \begin{array}{r} 12 \\ - 8 \\ \hline \end{array}\qquad \begin{array}{r} 6 \\ + 9 \\ \hline \end{array}\qquad \begin{array}{r} 14 \\ - 5 \\ \hline \end{array}$$

2.

$$\begin{array}{r} 10 \\ - 6 \\ \hline \end{array}\qquad \begin{array}{r} 7 \\ + 6 \\ \hline \end{array}\qquad \begin{array}{r} 5 \\ + 8 \\ \hline \end{array}\qquad \begin{array}{r} 11 \\ - 5 \\ \hline \end{array}\qquad \begin{array}{r} 11 \\ - 3 \\ \hline \end{array}\qquad \begin{array}{r} 7 \\ + 7 \\ \hline \end{array}\qquad \begin{array}{r} 13 \\ - 8 \\ \hline \end{array}$$

3.

$$\begin{array}{r} 10 \\ - 1 \\ \hline \end{array}\qquad \begin{array}{r} 4 \\ + 9 \\ \hline \end{array}\qquad \begin{array}{r} 7 \\ + 4 \\ \hline \end{array}\qquad \begin{array}{r} 11 \\ - 1 \\ \hline \end{array}\qquad \begin{array}{r} 8 \\ + 9 \\ \hline \end{array}\qquad \begin{array}{r} 12 \\ - 3 \\ \hline \end{array}\qquad \begin{array}{r} 7 \\ + 3 \\ \hline \end{array}$$

4.

$$\begin{array}{r} 6 \\ + 8 \\ \hline \end{array}\qquad \begin{array}{r} 11 \\ - 2 \\ \hline \end{array}\qquad \begin{array}{r} 9 \\ - 6 \\ \hline \end{array}\qquad \begin{array}{r} 8 \\ + 8 \\ \hline \end{array}\qquad \begin{array}{r} 8 \\ + 5 \\ \hline \end{array}\qquad \begin{array}{r} 18 \\ - 9 \\ \hline \end{array}\qquad \begin{array}{r} 14 \\ - 8 \\ \hline \end{array}$$

5.

$$\begin{array}{r} 8 \\ + 6 \\ \hline \end{array}\qquad \begin{array}{r} 15 \\ - 9 \\ \hline \end{array}\qquad \begin{array}{r} 8 \\ + 7 \\ \hline \end{array}\qquad \begin{array}{r} 11 \\ - 7 \\ \hline \end{array}\qquad \begin{array}{r} 10 \\ - 5 \\ \hline \end{array}\qquad \begin{array}{r} 5 \\ + 5 \\ \hline \end{array}\qquad \begin{array}{r} 9 \\ + 7 \\ \hline \end{array}$$

6.

$$\begin{array}{r} 13 \\ - 5 \\ \hline \end{array}\qquad \begin{array}{r} 9 \\ + 9 \\ \hline \end{array}\qquad \begin{array}{r} 17 \\ - 8 \\ \hline \end{array}\qquad \begin{array}{r} 8 \\ + 5 \\ \hline \end{array}\qquad \begin{array}{r} 8 \\ - 6 \\ \hline \end{array}\qquad \begin{array}{r} 16 \\ - 8 \\ \hline \end{array}\qquad \begin{array}{r} 10 \\ + 5 \\ \hline \end{array}$$

7.

$$\begin{array}{r} 5 \\ + 9 \\ \hline \end{array}\qquad \begin{array}{r} 15 \\ - 6 \\ \hline \end{array}\qquad \begin{array}{r} 17 \\ - 9 \\ \hline \end{array}\qquad \begin{array}{r} 7 \\ + 8 \\ \hline \end{array}\qquad \begin{array}{r} 3 \\ + 10 \\ \hline \end{array}\qquad \begin{array}{r} 14 \\ - 7 \\ \hline \end{array}\qquad \begin{array}{r} 16 \\ - 7 \\ \hline \end{array}$$

Name _____

Learn

268 children and 125 adults
gave to the Food Drive.
How many more children
than adults gave?
Subtract to find the difference.

You can use subtraction facts to help.

1 Subtract the ones.

hundreds	tens	ones
2	6	8
− 1	2	5
		3

2 Subtract the tens.

hundreds	tens	ones
2	6	8
− 1	2	5
	4	3

3 Subtract the hundreds.

hundreds	tens	ones
2	6	8
− 1	2	5
1	4	3

_143_ more children

Try it   Find each difference.

1.

hundreds	tens	ones
7	5	9
− 3	2	4
4	3	5

hundreds	tens	ones
6	4	7
−	1	3

hundreds	tens	ones
5	7	9
− 2	4	3

2.
$$573 - 240$$ $$826 - 702$$ $$694 - 51$$ $$748 - 223$$ $$937 - 6$$ $$436 - 126$$

Sum it Up

What subtraction facts can help you subtract 347 − 123?

Math at Home: Your child used basic facts to subtract hundreds, tens, and ones.
Activity: Have your child show you how to subtract 568 - 236.

Practice   Subtract.

3.

hundreds	tens	ones
6	5	7
− 2	4	1
4	1	6

hundreds	tens	ones
7	6	4
−	2	2

hundreds	tens	ones
7	5	4
− 3	4	2

4.

$$748 - 47 \qquad 625 - 413 \qquad 597 - 35 \qquad 687 - 563 \qquad 578 - 7 \qquad 484 - 24$$

5.

$$877 - 425 \qquad 678 - 53 \qquad 968 - 6 \qquad 476 - 324 \qquad 644 - 32 \qquad 709 - 427$$

6.

$$748 - 247 \qquad 625 - 13 \qquad 507 - 235 \qquad 687 - 63 \qquad 435 - 7 \qquad 484 - 124$$

7.

$$876 - 245 \qquad 568 - 43 \qquad 409 - 5 \qquad 629 - 314 \qquad 383 - 122 \qquad 836 - 25$$

Critical Thinking   Journal

8. What is the same about 489 and 498?
What is different?

Learn

Subtract 548 − 267.

1 Look at the ones.
Regroup if you
need to.
Subtract the ones

hundreds	tens	ones
☐	☐	☐
5	4	8
− 2	6	7
		1

2 Look at the tens.
Regroup if you
need to.
Subtract the tens.

hundreds	tens	ones
4	14	☐
5̸	4̸	8
− 2	6	7
	8	1

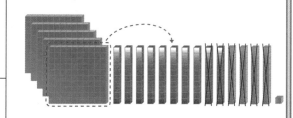

3 Subtract the
hundreds.

hundreds	tens	ones
4	14	☐
5̸	4̸	8
− 2	6	7
2	8	1

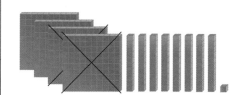

548 − 267 = 281

Try it   Find each difference. Use models.

1.

hundreds	tens	ones
3	15	☐
4̸	5̸	7
− 1	8	5
2	7	2

hundreds	tens	ones
☐	☐	☐
6	2	9
− 2	1	7

hundreds	tens	ones
☐	☐	☐
9	3	8
− 3	7	2

Sum it Up   How do you regroup to subtract tens?

 Math at Home: Your child subtracted 3-digit numbers deciding each time if regrouping was necessary.
Activity: Have your child show you how to subtract 537 − 291 and explain the regrouping to you.

Practice · Subtract. Use models.

2.

hundreds	tens	ones
[4] ~~5~~	[13] ~~3~~	[ ] 7
− 2	4	2
2	9	5

hundreds	tens	ones
[ ] 9	[ ] 2	[ ] 9
− 4	3	8

hundreds	tens	ones
[ ] 2	[ ] 7	[ ] 8
− 1	3	5

3.

hundreds	tens	ones
[ ] 6	[ ] 8	[ ] 7
− 2	9	5

hundreds	tens	ones
[ ] 4	[ ] 6	[ ] 5
− 3	9	4

hundreds	tens	ones
[ ] 7	[ ] 2	[ ] 4
− 3	6	2

4.

hundreds	tens	ones
[ ] 2	[ ] 5	[ ] 6
− 1	7	2

hundreds	tens	ones
[ ] 5	[ ] 3	[ ] 8
− 2	9	6

hundreds	tens	ones
[ ] 3	[ ] 4	[ ] 8
− 1	9	2

Problem Solving

5. How is subtracting 536 – 212 different from subtracting 536 – 292?

_____

536 – 212

536 – 292

Name _____

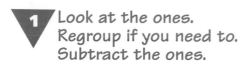

Sometimes you have to regroup ones and tens.

Find 436 - 278.

1 Look at the ones. Regroup if you need to. Subtract the ones.

hundreds	tens	ones
☐	②	16
4	3̷	6̷
− 2	7	8
		8

2 Look at the tens. Regroup if you need to. Subtract the tens.

hundreds	tens	ones
3	12	16
4̷	3̷	6̷
− 2	7	8
	5	8

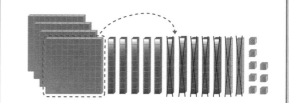

3 Subtract the hundreds.

hundreds	tens	ones
3	12	16
4̷	3̷	6
− 2	7	8
1	5	8

736 − 278 = __458__

Try it Subtract.

1.

hundreds	tens	ones
6	12	15
7̷	3̷	5̷
− 4	3	6
2	9	9

hundreds	tens	ones
☐	☐	☐
3	2	5
− 1	2	7

hundreds	tens	ones
☐	☐	☐
8	2	7
− 3	4	8

Sum it Up How do you know when to regroup?

Math at Home: Your child learned how to subtract 3-digit numbers with and without regrouping.
Activity: Have your child show you a subtraction problem that needs regrouping and a problem that does not need regrouping.

Practice   Subtract.

2.

hundreds	tens	ones
[6] 7̸	[12] 3̸	[15] 5̸
4	3	6
2	9	9

hundreds	tens	ones
☐ 3	☐ 2	☐ 5
1	2	7

hundreds	tens	ones
☐ 8	☐ 2	☐ 7
3	4	8

3.

```
  642      521      426      129      462      813
-  37    - 136    -  38    -  61    -   9    -  29
```

4.

```
  276      422      182      534      511      663
- 115    -  59    -   8    - 216    -  26    - 521
```

5.

```
  781      235      438      664      137      462
- 684    -  58    - 213    -  95    -   8    - 137
```

6.

```
  259      432      213      447      263      721
-  48    - 167    -   7    - 278    - 115    -  36
```

Problem Solving

Algebra & functions   Find each missing number.

7.

```
    723          ☐          ☐
  - ☐        - 258      - 468
  -----      -----      -----
    359        317        253
```

492   four hundred ninety-two

Name_____

Learn

The Friendship Club has 537 gifts to wrap. They wrap 311 presents on Friday and the same number on Saturday. Do all the gifts get wrapped?

$$537 - 311 = 226$$

226 is less than 311, so all the gifts will get wrapped. You can subtract nearest hundreds to see if your answer is reasonable.

Rewrite. Use nearest hundreds.

```
200    250    300 311  350    400    450    500 537 550   600
```

```
  537    nearest hundred →    500
- 311    nearest hundred →  - 300
  226                         200
```

226 is close to 200.
The answer is reasonable.

Try it   Subtract. Estimate to see if your answer is reasonable.

1.
```
  927          900          592
- 386        - 400        - 263          -
  541          500
```

2.
```
  483                       613
- 208          -          - 479          -
```

Sum it Up   How can you use estimation to check subtraction?

Math at Home: Your child estimated differences by finding the nearest hundreds.
Activity: Have your child subtract 497 - 329 and then estimate to see if the answer is reasonable.

four hundred ninety-three **493**

Subtract. Then find the nearest hundred.
Estimate each difference to check.

3.
$$\begin{array}{r} 673 \\ -\ 284 \\ \hline 389 \end{array}$$
$$\begin{array}{r} 700 \\ -\ 300 \\ \hline 400 \end{array}$$
$$\begin{array}{r} 789 \\ -\ 307 \\ \hline \end{array}$$
$$\begin{array}{r} \\ -\ \\ \hline \end{array}$$

4.
$$\begin{array}{r} 683 \\ -\ 485 \\ \hline \end{array}$$
$$\begin{array}{r} \\ -\ \\ \hline \end{array}$$
$$\begin{array}{r} 319 \\ -\ 126 \\ \hline \end{array}$$
$$\begin{array}{r} \\ -\ \\ \hline \end{array}$$

5.
$$\begin{array}{r} 529 \\ -\ 292 \\ \hline \end{array}$$
$$\begin{array}{r} \\ -\ \\ \hline \end{array}$$
$$\begin{array}{r} 202 \\ -\ 108 \\ \hline \end{array}$$
$$\begin{array}{r} \\ -\ \\ \hline \end{array}$$

6.
$$\begin{array}{r} 823 \\ -\ 385 \\ \hline \end{array}$$
$$\begin{array}{r} \\ -\ \\ \hline \end{array}$$
$$\begin{array}{r} 485 \\ -\ 219 \\ \hline \end{array}$$
$$\begin{array}{r} \\ -\ \\ \hline \end{array}$$

7.
$$\begin{array}{r} 599 \\ -\ 232 \\ \hline \end{array}$$
$$\begin{array}{r} \\ -\ \\ \hline \end{array}$$
$$\begin{array}{r} 213 \\ -\ 112 \\ \hline \end{array}$$
$$\begin{array}{r} \\ -\ \\ \hline \end{array}$$

Problem Solving

Mental Math

8. Roger and Cindy collected 448 magazines. They collected 248 magazines the first week. How many did they collect the second week?

_____ magazines

9. Emily and Julian made 220 cards on Monday and Tuesday. They made 110 cards on Monday. How many did they make on Tuesday?

_____ cards

Name _____

Learn

The students at Cleveland School had a toy sale. Jon sold a toy car for $1.75 and a radio for $3.49. What was the total cost of the toys?

$$\begin{array}{r} \$1.75 \\ +\ 3.49 \\ \hline \$5.24 \end{array}$$

The toys cost $5.24

Try it   Solve.

1. Sue buys a board game for $3.65. She has $5.00. How much change will she get?

_____

2. Larry buys a cassette tape for $1.75 and a back pack for $3.98. How much money does he spend altogether?

_____

Workspace

$$\begin{array}{r} \$5.00 \\ -\ 3.65 \\ \hline \$1.35 \end{array}$$

_____

Sum it Up   How do you add two money amounts?

Math at Home: Your child added and subtracted money amounts.
Activity: Have your child show you how to add $5.82 and $3.81.

3. Lisa sells a bear and a doll. How much money will she make?

_____

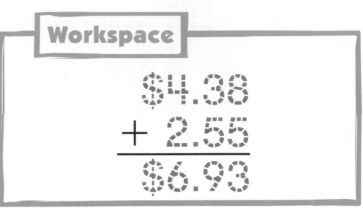

Workspace

$4.38
+ 2.55
$6.93

4. Tommy buys the toy truck. He has $4.00. How much change will he get?

_____

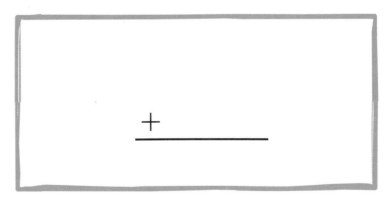

−
_____

5. You have $3.00. Can you buy a CD and the hat?

_____

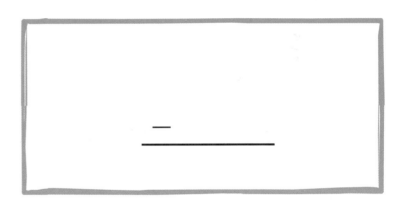

+
_____

6. What two items can you buy if you have $5.00? How much change will you get?

_____

Name _____

Plan a Food Drive

Your class wants to donate 650 pounds of food to the Main Street food bank. Here is the food that you have collected.

2nd Grade FOOD DRIVE

VEGETABLES 163 POUNDS

BREAD 147 POUNDS

CEREAL 275 POUNDS

RICE 212 POUNDS

You Decide!

Workspace

1. Choose three types of food to donate. Make a list. Do you have enough food?

   _____

   _____

   _____

   _____

2. How did you decide which foods to donate?

_____

_____

3. Draw a picture and write a problem to show the food that you donated.

_____

_____

_____

Workspace

4. **What if** the food bank needed 850 pounds of food? Would you have enough food if you donated everything that was collected? If not, how much more would you need to collect?

_____

_____

_____

_____

Name _____

How Much Water Does Your Class Use?

> **You Will Need**
> pencils

Water is a natural resource from the earth.

What to do

1. Find out how often your class uses water each day.

2. Whenever someone in your class uses water, make a tally mark.

3. At the end of the day, estimate how many times your class used water. Use your tallies to help you.

Water Use Chart

Day	Water Used
Monday	
Tuesday	
Wednesday	
Thursday	
Friday	
Total	

What did you find out?

1. What are some different ways your class uses water?

   _____

2. About how much water do you think your class uses in a week?

   _____

   _____

3. What did you notice about the amount of water your class uses?

   _____

Did You KNOW?

In many communities water is purified before you drink it.

Want to do more?

Think of ways that you can save water. Try them.

Name _____

Subtract.

1.
300	400	600	567	748
− 100	− 200	− 400	− 8	− 36

2.
587	446	538	476	212
− 32	− 225	− 163	− 329	− 8

Add or subtract.

3.
$7.00	$4.70	$8.77	$5.34	$6.25
− $2.00	+ $3.24	− $4.25	− $2.16	+ $1.16

Subtract. Estimate to see if your answer is reasonable.

4.
589		879	
− 212	− ____	− 221	− ____

5.
495		281	
− 286	− ____	− 107	− ____

Complete the bar graph to solve.

6. Vera collected 4 books for the Library Book Drive. Joey collected 2 books. Rory collected 10 books. Mrs. Parkins gave them 4 more books apiece. How many books did each collect?

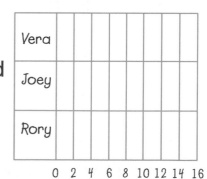

_____

Name _____

It All Adds Up!

- Work with a partner. Roll the **2** 6 times.

- Put each number in one place in the addition or the subtraction problem below.

- Find the sum or difference.

- The player with the answer closest to 500 wins.

- Play again.

You Will Need

2

hundreds	tens	ones
□	□	□
□	□	□
□	□	□

+

hundreds	tens	ones
□	□	□
□	□	□
□	□	□

−

Chapter Review

Name _____

Language and Math

Complete. Use a word from the list.

1. To find the difference of 569 − 384

   you must _____.

2. In 225 + 134 = 359, the number 225

   is an _____.

Read these words.

Math Words

addend
difference
regroup

Concepts and Skills

Add.

3.
```
  225        386       $4.32        647        408
+ 319      + 213      + $1.23     + 189      +  59
```

4.
```
  369       $4.26        513        126       $3.62
+ 235      + $1.44     + 235      + 786      + $3.42
```

5.
```
  300        418       $7.00        662        528
+ 600      +  39      + $2.00     + 139      +   7
```

McGraw-Hill School Division

Subtract.

6.
$$\begin{array}{r} 562 \\ - 231 \\ \hline \end{array}$$
$$\begin{array}{r} \$4.95 \\ - \$1.32 \\ \hline \end{array}$$
$$\begin{array}{r} 227 \\ - 128 \\ \hline \end{array}$$
$$\begin{array}{r} 431 \\ - 207 \\ \hline \end{array}$$
$$\begin{array}{r} 813 \\ - 494 \\ \hline \end{array}$$

7.
$$\begin{array}{r} 458 \\ - 29 \\ \hline \end{array}$$
$$\begin{array}{r} 612 \\ - 9 \\ \hline \end{array}$$
$$\begin{array}{r} 254 \\ - 116 \\ \hline \end{array}$$
$$\begin{array}{r} \$4.81 \\ - \$2.69 \\ \hline \end{array}$$
$$\begin{array}{r} 952 \\ - 18 \\ \hline \end{array}$$

8.
$$\begin{array}{r} 473 \\ - 289 \\ \hline \end{array}$$
$$\begin{array}{r} 756 \\ - 47 \\ \hline \end{array}$$
$$\begin{array}{r} \$5.47 \\ - \$1.25 \\ \hline \end{array}$$
$$\begin{array}{r} 384 \\ - 216 \\ \hline \end{array}$$
$$\begin{array}{r} 845 \\ - 269 \\ \hline \end{array}$$

Problem Solving

Complete the bar graph to solve.

9. On Monday, Larry walked for 8 minutes. Rob walked for 4 minutes. Phil walked for 2 minutes. On Tuesday, they each walked 2 minutes more. How many minutes did each of them walk?

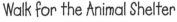

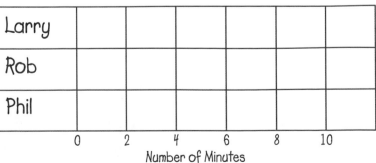

Walk for the Animal Shelter

Name

| Larry |
| Rob |
| Phil |

0   2   4   6   8   10
Number of Minutes

_____

Extra Practice

Name _____

Add or subtract. Use the letters to find the secret message.

378 − 24	367 − 146	146 − 26	611 − 125	143 − 7	603 − 241
D	**U**	**Y**	**S**	**J**	**T**

789 − 461	259 − 58	575 − 163	443 − 216	430 − 315	512 − 436
R	**F**	**C**	**Q**	**E**	**V**

213 + 143	217 + 239	419 − 205	340 + 220	433 − 6	304 + 111
O	**B**	**K**	**A**	**I**	**N**

414 − 238	200 + 400	569 − 4	560 + 250	228 − 103	345 + 110
M	**H**	**G**	**L**	**W**	**P**

Secret Message:

Y __ __ __ __ __ __ __ __
120   356   221   412   560   415   456   115   560

__ __ __ __ __ __   __ __ __ !
600   115   810   455   115   328   362   356   356

Each sum is 15.

The sum of the numbers for each row, column, and diagonal in the magic square is the same.

$4 + 3 + 8 = \underline{15}$ $2 + 7 + 6 = \underline{15}$

4	9	2
3	5	7
8	1	6

$4 + 9 + 2 = \underline{15}$

$3 + 5 + 7 = \underline{15}$

$8 + 1 + 6 = \underline{15}$

$8 + 5 + 2 = \underline{15}$ $9 + 5 + 1 = \underline{15}$ $4 + 5 + 6 = \underline{15}$

Find the magic sum for each magic square.

1.

5	10	3
4	6	8
9	2	7

8	18	4
6	10	14
16	2	12

27	34	29
32	30	28
31	26	33

_____   _____   _____

Find the missing numbers for each magic square.

2.

12	27	
	15	21
24	3	18

30		60
105	75	45
	15	120

	129	
43	215	387
258		86

Name _____

Add.

1. | 300 | 413 | 179 | 262 | 441 | $4.10 |
    |---|---|---|---|---|---|
    | + 200 | + 234 | + 438 | + 19 | + 328 | +$1.68 |

2. | 583 | 261 | 457 | 300 | 219 | $4.56 |
    |---|---|---|---|---|---|
    | + 415 | + 49 | + 228 | + 200 | + 642 | + $1.23 |

Subtract.

3. | 468 | 289 | 325 | 189 | 642 | $2.76 |
    |---|---|---|---|---|---|
    | − 213 | − 125 | − 46 | − 8 | − 396 | − $1.34 |

4. | 657 | 176 | 449 | 234 | 564 | $4.89 |
    |---|---|---|---|---|---|
    | − 89 | − 9 | − 261 | − 125 | − 29 | − $1.25 |

Complete the bar graph to solve.

5. Tammy read for 15 minutes before school. Robbie read for 10 minutes. Jill read for 20 minutes. If each of them read for 10 more minutes at lunch time, how many minutes did each of them read in all?

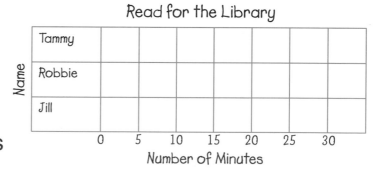

Read for the Library

_____

You have $8.50.
Which two items can you buy?
Show as many combinations as you can.
Show the sums.

_____

_____

_____

_____

_____

Name _____

Choose the correct answer.

Statistics, Data Analysis, and Probability

Use the chart.

Favorite Sport	
Soccer	⚽⚽⚽⚽
Baseball	⚾⚾
Basketball	🏀🏀🏀
Football	🏈🏈

1. Which sport got 4 votes?

   ○ Soccer

   ○ Baseball

   ○ Basketball

   ○ Football

2. The data shows the number of children in each of the five Second Grade classes. What is the range in the number of children?

   25, 26, 27, 29, 29

   ○ 4

   ○ 5

   ○ 25

   ○ 29

3. What could the next number be in the pattern?

   1 2 3 3 1 2 3

   ○ 1

   ○ 2

   ○ 3

   ○ 4

Mathematical Reasoning

4. Joe is behind Kelly in line. Kelly is between Lou and Maria. Lou is first in line. List the people in line in order. Tell how you found out.

   _____

   _____

5. Carl had $4.25. He spent $2.60. How much money did he have left?

   ○ $6.85

   ○ $1.65

   ○ $1.60

   ○ $1.55

   Explain how you found your answer.

   _____

6. Betsy has $4.50. She gives her brother 50 cents. How much does she have left?

   ○ $4.25

   ○ $4.00

   ○ $4.05

   ○ $3.90

Add.

Number Sense

7. Add.

$$327 + 468$$

- ◯ 765
- ◯ 775
- ◯ 785
- ◯ 795

8. Subtract.

$$617 - 234$$

- ◯ 283
- ◯ 383
- ◯ 483
- ◯ 851

9. Which shows the number in expanded form?

- ◯ 5 + 2 + 8
- ◯ 50 + 20 + 8
- ◯ 500 + 20 + 8
- ◯ 500 + 200 + 8

Algebra and Functions

10. Which is the turnaround fact for 5 + 9 = 14?

- ◯ 5 + 5 = 10
- ◯ 9 + 5 = 14
- ◯ 9 + 9 = 18
- ◯ 14 - 7 = 7

11. Which number makes this sentence true?

$$10 + \boxed{} = 24$$

- ◯ 6
- ◯ 10
- ◯ 14
- ◯ 34

12. Dan had 8 pencils. He gave 2 away. How many did he have left?

Which number sentence would you use to solve this problem?

- ◯ 8 - 2 = 6
- ◯ 8 + 2 = 10
- ◯ 10 - 2 = 8
- ◯ 2 + 6 = 8

theme

Coast to Coast

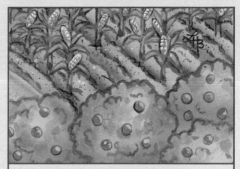

Use the Data

What number stories can you tell about the picture?

What You Will Learn

In this chapter you will learn how to:

- Find the product given groups of even numbers.

- Find how many groups and how many in each group.

- Find how many in each group with remainders.

- Draw a picture to solve problems.

Susan's Farm

Story by Ay-Jinan Han • Illustrated by Linda Howard Bittner

Dear Family,

In Chapter 14, I will multiply and divide. Here are new vocabulary words and an activity that we can do together.

Fair Shares

- Put 12 straws on the table.

- Ask your child to find different ways to make equal groups of straws.
- You can repeat this activity with other numbers, such as 14 or 18.

use

colorful straws

Math Words

multiplication sentence

$2 \times 4 = 8$

$2 \times 3 = 6$ ← product

↑ ↑
factors

division sentence

$10 \div 2 = 5$

Additional activities at
www.mhschool.com/math

Learn

The crates are in equal groups.

You can skip count to find how many.

2 , 4 , 6 , 8

There are ___8___ crates in all.

Try it   Skip count. Write how many in all.

1.

_5_   _10_   ____   ____   ____ in all

2.

____   ____   ____   ____   ____ in all

Sum it Up   When can you skip count to find how many in all?

 Math at Home: Your child learned to skip count to find how many in all.
Activity: Ask your child to skip count by 5 and to use beans to remember how many times he or she counted 5.

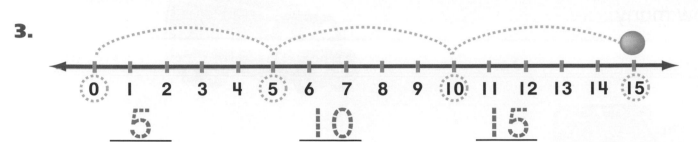

You can use a number line to help you skip count.

3.

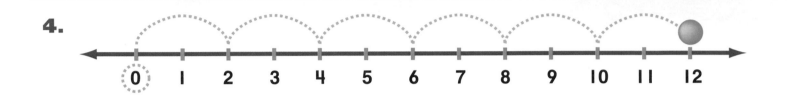

5   10   15

4.

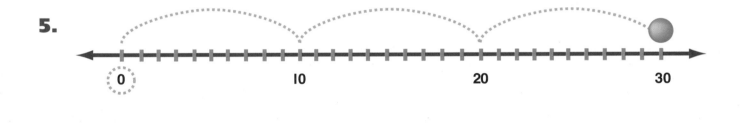

____   ____   ____   ____   ____   ____

5.

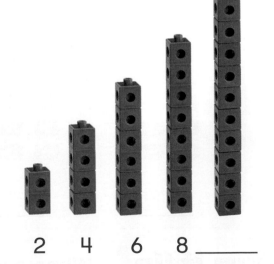

0   10   20   30

____   ____   ____

Critical Thinking

6. Look at the picture.

What could the next number be?

What could the pattern be? Explain.

2   4   6   8 ____

Name _____

Learn

When groups are equal you can use repeated addition or multiplication to find how many in all.

Use . Make 5 groups of 2s. How many are there in all?

Math Words
repeated addition
multiplication
multiplication
sentence

I can write a multiplication sentence.
$5 \times 2 = 10$

I can write an addition sentence.
$2 + 2 + 2 + 2 + 2 = 10$

$\underline{2} + \underline{2} + \underline{2} + \underline{2} + \underline{2} = \underline{10}$

$\underline{5} \times \underline{2} = \underline{10}$

Try it

Use  to make equal groups. Find how many in all.

1. Make 2 groups of 4s.

_____8_____ in all

2. Make 10 groups of 3s.

_____ in all

Workspace

 **Sum it Up**   How would you write a multiplication sentence for $3 + 3 + 3$?

 Math at Home: Your child learned how repeated addition and multiplication are related.
Activity: Ask your child to explain to you what multiplication is and how it is used when there are equal numbers of things in groups.

five hundred fifteen   **515**

Repeated addition and multiplication are related.

3.

$$\underline{4} + \underline{4} = \underline{8}$$

$$\underline{2} \times \underline{4} = \underline{8}$$

4.

$$\underline{\hphantom{0}} + \underline{\hphantom{0}} + \underline{\hphantom{0}} + \underline{\hphantom{0}} + \underline{\hphantom{0}} = \underline{\hphantom{0}}$$

$$\underline{\hphantom{0}} \times \underline{\hphantom{0}} = \underline{\hphantom{0}}$$

5.

$$\underline{\hphantom{0}} + \underline{\hphantom{0}} = \underline{\hphantom{0}}$$

$$\underline{\hphantom{0}} \times \underline{\hphantom{0}} = \underline{\hphantom{0}}$$

6.

$$\underline{\hphantom{0}} + \underline{\hphantom{0}} + \underline{\hphantom{0}} + \underline{\hphantom{0}} + \underline{\hphantom{0}} = \underline{\hphantom{0}}$$

$$\underline{\hphantom{0}} \times \underline{\hphantom{0}} = \underline{\hphantom{0}}$$

Problem Solving **Algebra & functions**

Use the addition sentence to complete the multiplication sentence.

7. $4 + 4 + 4 + 4 + 4 = 20$ $\boxed{} \times 4 = 20$

Learn

You can write a multiplication sentence to show an **array**.

I can count the rows. There are 4 rows. There are 2 balls in each row.

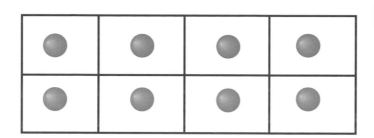

2 rows $\times$ 4 in each row

$2 \times 4 = 8$

Try it

Write a multiplication sentence to show each array.

1.

$\underline{2} \times \underline{6} = \underline{12}$

rows   in each row   in all

2.

$\underline{} \times \underline{} = \underline{}$

rows   in each row   in all

3.

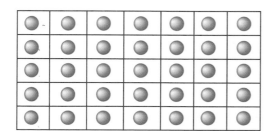

$\underline{} \times \underline{} = \underline{}$

rows   in each row   in all

4.

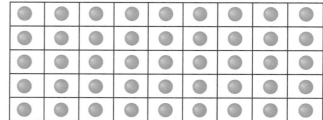

$\underline{} \times \underline{} = \underline{}$

rows   in each row   in all

Sum it Up

How does counting the number of rows and how many in each row help you write a multiplication sentence?

Math at Home: Your child practiced writing multiplication sentences using multiplication arrays. **Activity:** Write an addition sentence, such as 3 + 3 + 3 = 9. Have your child write a multiplication sentence to show the addition sentence.

five hundred seventeen **517**

Write a multiplication sentence to show each array.

5.

_____

6.

_____

7.

_____

Find each product.

8. $5 \times 2 =$ _____ $10 \times 2 =$ _____ $5 \times 4 =$ _____

9. $10 \times 5 =$ _____ $2 \times 4 =$ _____ $5 \times 3 =$ _____

Problem Solving

Draw a picture to solve.

10. Is 2×5 the same as 5×2?

_____

Explain. Then draw a picture
to show why or why not.

_____

Workspace

Learn

You can multiply across or down.

I can multiply across.
$2 \times 3 = 6$

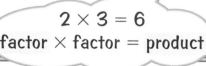

I can multiply down.
$$\begin{array}{r} 3 \\ \times\, 2 \\ \hline 6 \end{array}$$

$2 \times 3 = 6$
factor $\times$ factor = product

Try it   Find each product.

1.

$$\begin{array}{r} 5 \\ \times\, 2 \\ \hline 10 \end{array}$$

$2 \times 5 = $ ____

$$\begin{array}{r} 5 \\ \times\, 3 \\ \hline \end{array}$$

$3 \times 5 = $ ____

2.

$$\begin{array}{r} 3 \\ \times\, 10 \\ \hline \end{array}$$

$10 \times 3 = $ ____   ____

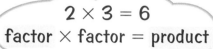

$$\begin{array}{r} 6 \\ \times\, 2 \\ \hline \end{array}$$

$2 \times 6 = $ ____   ____

Sum it Up

Why is the product the same whether you multiply
across or down?

 Math at Home: Your child multiplied the same two numbers across and down.
Activity: Give your child a multiplication problem such as 3 x 2. Ask your
child to write the multiplication across, then down, finding the product
each time.

five hundred nineteen   **519**

 When you multiply across or down, the product is the same.

3.

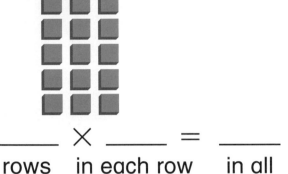

$$\underline{2} \times \underline{4} = \underline{8}$$
rows   in each row   in all

$$\underline{} \times \underline{} = \underline{}$$
rows   in each row   in all

4.

$$\underline{} \times \underline{} = \underline{}$$
rows   in each row   in all

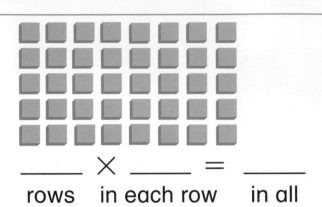

$$\underline{} \times \underline{} = \underline{}$$
rows   in each row   in all

Find each product.

5. $2 \times 6 = \underline{}$ $5 \times 5 = \underline{}$ $10 \times 6 = \underline{}$

6. $2 \times 8 = \underline{}$ $10 \times 7 = \underline{}$ $2 \times 9 = \underline{}$

7.
$$\begin{array}{r} 3 \\ \times 5 \\ \hline \end{array} \qquad \begin{array}{r} 4 \\ \times 2 \\ \hline \end{array} \qquad \begin{array}{r} 2 \\ \times 5 \\ \hline \end{array} \qquad \begin{array}{r} 4 \\ \times 5 \\ \hline \end{array} \qquad \begin{array}{r} 10 \\ \times 2 \\ \hline \end{array} \qquad \begin{array}{r} 10 \\ \times 5 \\ \hline \end{array}$$

8.
$$\begin{array}{r} 3 \\ \times 2 \\ \hline \end{array} \qquad \begin{array}{r} 3 \\ \times 5 \\ \hline \end{array} \qquad \begin{array}{r} 6 \\ \times 2 \\ \hline \end{array} \qquad \begin{array}{r} 3 \\ \times 5 \\ \hline \end{array} \qquad \begin{array}{r} 6 \\ \times 5 \\ \hline \end{array} \qquad \begin{array}{r} 7 \\ \times 2 \\ \hline \end{array}$$

9. Does 5×10 have the same product as 10×5? $\underline{}$

Explain why or why not.

_____

Learn

You can multiply factors in any order.

$$2 \times 4 = 8 \qquad\qquad 4 \times 2 = 8$$

Try it Find each product.

1. 5 groups of 4

$$5 \times 4 = \underline{20}$$

4 groups of 5

$$4 \times 5 = \underline{20}$$

2. 2 groups of 5

$$2 \times 5 = \underline{}$$

5 groups of 2

$$5 \times 2 = \underline{}$$

 Does changing the order of the factors change what the product will be? Explain.

Math at Home: Your child multiplied factors in any order.
Activity: Ask your child to write two multiplication sentences for the following: 5 trains, each with 4 cars.

3. 2 groups of 6   6 groups of 2

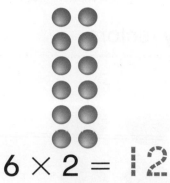

$2 \times 6 = 12$ $6 \times 2 = 12$

4. 5 groups of 4   4 groups of 5

$5 \times 4 = \underline{}$ $4 \times 5 = \underline{}$

5.
$\begin{array}{r} 4 \\ \times 5 \\ \hline \end{array}$
$\begin{array}{r} 5 \\ \times 4 \\ \hline \end{array}$
$\begin{array}{r} 6 \\ \times 5 \\ \hline \end{array}$
$\begin{array}{r} 5 \\ \times 6 \\ \hline \end{array}$
$\begin{array}{r} 10 \\ \times 5 \\ \hline \end{array}$
$\begin{array}{r} 5 \\ \times 10 \\ \hline \end{array}$

6. $2 \times 7 = \underline{}$ $7 \times 2 = \underline{}$

Spiral Review and Test Prep

Choose the correct answer.

7. $5 \times 8 = \boxed{}$

- ⬭ 3
- ⬭ 13
- ⬭ 40
- ⬭ 58

8. Which number makes this sentence true?

quarter hour = $\boxed{}$ minutes

- ⬭ 10
- ⬭ 15
- ⬭ 30
- ⬭ 60

Multiply. Complete the multiplication table.

1.

X	0	1	2	3	4	5	6	7	8	9	10
2	0	2	4	6							
5	0	5	10	15							
10	0	10	20	30							

Multiply.

2. $2 \times 2 =$ _____

3. $2 \times 8 =$ _____

4. $5 \times 5 =$ _____

5. $5 \times 7 =$ _____

6. $10 \times 0 =$ _____

7. $10 \times 9 =$ _____

8. $2 \times 9 =$ _____

9. $5 \times 8 =$ _____

10. $10 \times 6 =$ _____

11. $5 \times 4 =$ _____

12. $2 \times 7 =$ _____

13. $10 \times 10 =$ _____

Math at Home: Your child practiced finding products using the multiplication tables of 2s, 5s, and 10s. **Activity:** Have your child tell you how he or she can find the product of a number multiplied by 10.

five hundred twenty-three **523**

Facts Practice: Multiplication

Multiply.

1.
$$\begin{array}{r} 8 \\ \times 5 \\ \hline \end{array} \qquad \begin{array}{r} 5 \\ \times 2 \\ \hline \end{array} \qquad \begin{array}{r} 3 \\ \times 5 \\ \hline \end{array} \qquad \begin{array}{r} 7 \\ \times 5 \\ \hline \end{array} \qquad \begin{array}{r} 5 \\ \times 5 \\ \hline \end{array} \qquad \begin{array}{r} 9 \\ \times 5 \\ \hline \end{array} \qquad \begin{array}{r} 3 \\ \times 10 \\ \hline \end{array}$$

2.
$$\begin{array}{r} 4 \\ \times 2 \\ \hline \end{array} \qquad \begin{array}{r} 9 \\ \times 2 \\ \hline \end{array} \qquad \begin{array}{r} 2 \\ \times 2 \\ \hline \end{array} \qquad \begin{array}{r} 6 \\ \times 2 \\ \hline \end{array} \qquad \begin{array}{r} 3 \\ \times 2 \\ \hline \end{array} \qquad \begin{array}{r} 2 \\ \times 10 \\ \hline \end{array} \qquad \begin{array}{r} 2 \\ \times 8 \\ \hline \end{array}$$

3.
$$\begin{array}{r} 5 \\ \times 1 \\ \hline \end{array} \qquad \begin{array}{r} 5 \\ \times 4 \\ \hline \end{array} \qquad \begin{array}{r} 5 \\ \times 10 \\ \hline \end{array} \qquad \begin{array}{r} 5 \\ \times 6 \\ \hline \end{array} \qquad \begin{array}{r} 2 \\ \times 1 \\ \hline \end{array} \qquad \begin{array}{r} 2 \\ \times 8 \\ \hline \end{array} \qquad \begin{array}{r} 9 \\ \times 2 \\ \hline \end{array}$$

4.
$$\begin{array}{r} 2 \\ \times 5 \\ \hline \end{array} \qquad \begin{array}{r} 2 \\ \times 7 \\ \hline \end{array} \qquad \begin{array}{r} 10 \\ \times 1 \\ \hline \end{array} \qquad \begin{array}{r} 10 \\ \times 3 \\ \hline \end{array} \qquad \begin{array}{r} 10 \\ \times 8 \\ \hline \end{array} \qquad \begin{array}{r} 10 \\ \times 4 \\ \hline \end{array} \qquad \begin{array}{r} 10 \\ \times 9 \\ \hline \end{array}$$

5.
$$\begin{array}{r} 2 \\ \times 10 \\ \hline \end{array} \quad \begin{array}{r} 5 \\ \times 10 \\ \hline \end{array} \quad \begin{array}{r} 7 \\ \times 10 \\ \hline \end{array} \quad \begin{array}{r} 6 \\ \times 10 \\ \hline \end{array} \quad \begin{array}{r} 9 \\ \times 10 \\ \hline \end{array} \quad \begin{array}{r} 10 \\ \times 10 \\ \hline \end{array} \quad \begin{array}{r} 10 \\ \times 3 \\ \hline \end{array}$$

6.
$$\begin{array}{r} 4 \\ \times 2 \\ \hline \end{array} \qquad \begin{array}{r} 4 \\ \times 5 \\ \hline \end{array} \qquad \begin{array}{r} 3 \\ \times 2 \\ \hline \end{array} \qquad \begin{array}{r} 3 \\ \times 5 \\ \hline \end{array} \qquad \begin{array}{r} 3 \\ \times 10 \\ \hline \end{array} \qquad \begin{array}{r} 4 \\ \times 10 \\ \hline \end{array} \qquad \begin{array}{r} 4 \\ \times 7 \\ \hline \end{array}$$

7.
$$\begin{array}{r} 5 \\ \times 6 \\ \hline \end{array} \qquad \begin{array}{r} 2 \\ \times 6 \\ \hline \end{array} \qquad \begin{array}{r} 1 \\ \times 5 \\ \hline \end{array} \qquad \begin{array}{r} 2 \\ \times 7 \\ \hline \end{array} \qquad \begin{array}{r} 10 \\ \times 1 \\ \hline \end{array} \qquad \begin{array}{r} 8 \\ \times 5 \\ \hline \end{array} \qquad \begin{array}{r} 4 \\ \times 4 \\ \hline \end{array}$$

Name _____

Make a Prediction

Reading Skill   Sometimes you can make a prediction to solve a problem.

The train has 10 cars. The first car is carrying 10 boats. Each car carries the same number of boats. Predict how many boats the train is carrying.

1. What do you want to find out? _____

2. What do you know? _____

3. How many cars are in the train? _____

   Which sentence of the problem tells you this? _____

4. Predict the number of boats there are on each train car. _____ .

5. How can you predict how many boats in all? _____

   _____ cars x _____ boats in each car = _____ boats

 Math at Home: Your child read to make predictions.
Activity: Ask your child to explain why his or her answer to question 5 is a prediction.

5 planes are waiting to take off.
The first plane has 4 engines.
Each plane has the same number of engines.
Predict how many engines there are altogether.

6. What do you want to find out?

_____

7. What do you know?

_____

8. How many planes are waiting to take off? _____

9. Predict the number of engines each plane has. _____

Explain your prediction.

_____

10. How can you find how many engines in all?

_____

_____ planes × each with _____ engines = _____ engines

11. Use the picture. Write a problem.

Name _____

Skip count by 5s. Write the numbers.

1.

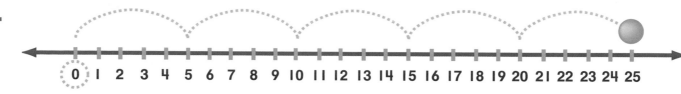

____ ____ ____ ____ ____

Add. Then multiply.

2. $5 + 5 + 5 =$ ____   ____ $\times$ ____ $=$ ____

Write a multiplication sentence to show each array.

3.

   _____

4.

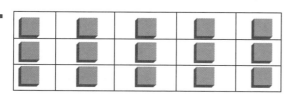   _____

Find each product.

5. $2 \times 5 =$ —— $5 \times 8 =$ —— $5 \times 4 =$ ——

6. $10 \times 3 =$ —— $2 \times 9 =$ —— $5 \times 6 =$ ——

7. $2 \times 7 =$ —— $5 \times 5 =$ —— $10 \times 7 =$ ——

8.
$$\begin{array}{cccccc} 5 & 2 & 5 & 5 & 10 & 2 \\ \times 3 & \times 6 & \times 7 & \times 9 & \times 4 & \times 10 \end{array}$$

Add.

1.
$$
\begin{array}{r} 28 \\ + 17 \end{array}
\qquad
\begin{array}{r} 36 \\ + 48 \end{array}
\qquad
\begin{array}{r} 12 \\ + 45 \end{array}
\qquad
\begin{array}{r} 40 \\ + 37 \end{array}
\qquad
\begin{array}{r} 39 \\ + 58 \end{array}
\qquad
\begin{array}{r} 60 \\ + 23 \end{array}
$$

Subtract.

2.
$$
\begin{array}{r} 93 \\ - 52 \end{array}
\qquad
\begin{array}{r} 77 \\ - 32 \end{array}
\qquad
\begin{array}{r} 49 \\ - 18 \end{array}
\qquad
\begin{array}{r} 63 \\ - 49 \end{array}
\qquad
\begin{array}{r} 81 \\ - 68 \end{array}
\qquad
\begin{array}{r} 60 \\ - 25 \end{array}
$$

Write the time
under each clock.

3.

_____   _____

 **TECHNOLOGY LINK**

Model Multiplication

- Use counters.
- Choose a mat to show one number.
- Stamp out 4 rows of 3 frogs.
- What multiplication fact is shown?

_____

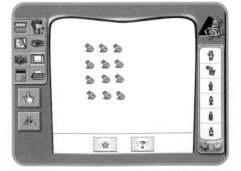

1. Use counters. Stamp out 5 rows of 2. What multiplication fact is shown? _____

2. Stamp out other multiplication facts.

For more practice use Math Traveler.™

Name _____

Draw a Picture

What do you know?

Read → The farmers load 2 crates of corn on 3 trucks. How many crates of corn are on the trucks?

What do you need to find out?

Plan → Draw a picture.

Solve → $\underline{2} \times \underline{3} = \underline{6}$

Look Back → Does your answer make sense? Why?

Solve. Draw a picture.

1. 2 trucks deliver 2 cartons each of eggs to the hospital. How many cartons are there in all? _____ cartons

2. 2 trucks deliver 5 crates each of milk to the school. How many crates are there in all? _____ crates

Sum it Up How can drawing a picture help you solve the problem?

Math at Home: Your child practiced solving problems by drawing a picture.
Activity: There are 2 trucks bringing 8 crates each of apples to the market. Ask your child how to find the number of crates in all.

3. 10 trucks each delivered 5 crates of carrots to the market. How many crates are there in all?

_____ crates

4. Each day the baker delivers 5 baskets of bread to the restaurant. Each basket holds 5 loaves of bread. How much bread does the restaurant get in all?

_____ loaves of bread

5. 2 trucks each carried 7 boxes of corn to the restaurant. How many boxes of corn are there in all?

_____ boxes

6. Write your own problem about how a farmer delivered bananas to the market. Draw a picture to show your story.

_____

Name _____

Learn

You can subtract to find the number of equal groups.

Math Words

equal groups

Use .
Separate 12 cubes into equal groups of 4 each.
How many equal groups of 4 can you make?

You can use repeated subtraction to find the number of equal groups.

$12 - 4 = 8$ $8 - 4 = 4$ $4 - 4 = 0$

I made 3 groups of 4.

Try it

Use .
How many equal groups can you make?

1. Use 6 .

How many equal groups can you make?

Subtract groups of 3.

You get __2__ groups of 3.

Use 15 .

Subtract groups of 5.

You get _____ groups of 5.

2. Use 10 .

Subtract groups of 2.

You get _____ groups of 2.

Use 12 .

Subtract groups of 3.

You get _____ groups of 3.

Sum it Up

How does subtracting equal groups help you divide?

Math at Home: Your child practiced doing repeated subtraction to divide into equal groups.
Activity: Put 20 peanuts or crackers on the table. Ask your child to share them equally in groups of 5.

Practice

Use .

How many equal groups can you make?

Remember, each equal group has the same number of cubes.

3. Use 12 .

Subtract groups of 6.

You get __2__ groups of 6.

Use 16 .

Subtract groups of 4.

You get _____ groups of 4.

---

4. Use 21 .

Subtract groups of 7.

You get _____ groups of 7.

Use 21 .

Subtract groups of 3.

You get _____ groups of 3.

---

5. Use 30 .

Subtract groups of 5.

You get _____ groups of 5.

Use 18 .

Subtract groups of 2.

You get _____ groups of 2.

Problem Solving

6. The rancher moves 12 horses to another field.

Each horse trailer holds 2 horses.

How many trailers will it take?

_____ trailers

Name _____

Learn

Math Words

divide
division sentence
quotient

Each truck carries 5 horses. How many trucks do you need to carry the horses?

You can subtract or divide to find the answer.

10 − 5 = 5   5 − 5 = 0

How many times did you subtract? __2__ times

10 ÷ 5 = 2 trucks

It takes __2__ trucks.

> I can write a division sentence. The answer is the quotient.

Try it   Subtract. Then divide.

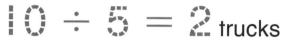

1. There are 20 trucks to load onto

   train cars. Each train car holds 5 trucks.

   How many times can you

   subtract 5? __4__

   20 ÷ 5 = __4__

   How many train
   cars do you need? __4__ train cars

   How does counting how many times you can subtract help you divide?

      Math at Home: Your child practiced relating division and repeated subtraction.
Activity: Ask your child to show you how to find 16 ÷ 3 using subtraction.

five hundred thirty-three   **533**

Subtract. Then divide.

2. There are 12 cars to load onto a big truck.
Each truck can carry 3 cars.

How many times can you subtract 3? __4__

12 ÷ 3 = _____

How many trucks do you need? _____ trucks

Draw a picture. Solve.
Use repeated subtraction. Divide.

3. 16 people

4 in each bus

16 ÷ 4 = _____

How many buses do you need?

_____ buses

4. 20 pigs

4 in each truck

20 ÷ 4 = _____

How many trucks do you need? _____ trucks

Spiral Review and Test Prep

5. Sally packs oranges. She has
20 oranges to pack. Each
package holds 5 oranges. How
many packages does she need?

- ◯ 25
- ◯ 15
- ◯ 5
- ◯ 4

6. Which number completes the
addition?

☐ + 30 = 60

- ◯ 70
- ◯ 30
- ◯ 50
- ◯ 40

Learn

You can divide to make equal groups.

Math Words

fair share

Let's divide the crayons equally between us.

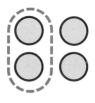

$6 \div 2 = 3$

We each have ___3___ crayons.
We each have a fair share.

Try it

Make equal groups.
Write how many in each group.

1. 4 trucks

2 groups of people

___2___ in each group

2. 10 trains

2 groups of people

_____ in each group

3. 8 planes

4 groups of people

_____ in each group

Sum it Up

Can you make equal groups of 7 toys between 2 children? Explain.

 **Math at Home:** Your child practiced making equal groups.
Activity: Ask your child to make 3 equal groups of a set of plates.

Practice Make equal groups. Divide. Write how many in each group.

Remember, divide to make equal groups.

4. 6 counters

3 groups

$6 \div 3 = \underline{2}$

$\underline{2}$ in each group

5. 8 counters

2 groups

$8 \div 2 = \underline{}$

$\underline{}$ in each group

6. 10 counters

5 groups

$10 \div 5 = \underline{}$

$\underline{}$ in each group

Circle Fair or Not Fair.

7. There are 10 apples to share among 5 friends. Will takes 3 apples.

Fair   Not Fair

8. There are 6 cows to load on 2 trucks. The farmer puts 4 cows on one truck.

Fair   Not Fair

9. There are 15 trees to load on 5 trucks. The farmer puts 3 trees on one truck.

Fair   Not Fair

Problem Solving

Number Sense

10. The farmer picks 11 crates of peaches. There are 2 trucks to carry them. Can each van carry the same number of crates of peaches?

_____

Explain why.

_____

Name _____

Learn

Sometimes when you divide, you have a **remainder**.

Pete has 14 crayons. He makes 3 equal groups. There are 4 in each group. Pete has 2 crayons left over.

Leftovers are called remainders.

Math Words

remainder

14 crayons in all
3 equal groups

<u>4</u> crayons in each group

<u>2</u> crayons left over

3 equal groups   Remainder

Try it

Make equal groups.
Write how many are in each group
and how many are left over.

1. 16 pumpkins

3 trucks

_____ pumpkins in each truck

_____ pumpkins left over

Workspace

2. 18 cows

4 trucks

_____ cows in each truck

_____ cows left over

Sum it Up   You want to divide your crayons into equal groups of 5. Can there be a remainder that is greater than 5? Explain.

 Math at Home: Your child divided with remainders.
Activity: Ask your child to explain to you how to divide 12 by
5. Ask what another name is for leftovers.

five hundred thirty-seven **537**

Make equal groups. Write how many are in each group and how many are left over.

3. 9 train cars
   2 engines

_____ train cars in each group

_____ train car left over

4. 15 people
   7 tandem bicycles

_____ people on each tandem bicycle

_____ person left over

Divide into equal groups. Write how many in each group. Write how many are left over.

5. 12 into 3 equal groups

_____ in each group   _____ left over

6. 12 into 4 equal groups

_____ in each group   _____ left over

7. 12 into 5 equal groups

_____ in each group   _____ left over

8. 12 into 6 equal groups

_____ in each group _____ left over

9. The farmer has 62 apples to pack. She can pack them into small or large crates. Which sentence is correct?

◯ The farmer needs 62 crates.

◯ The farmer needs fewer crates if she uses large crates.

Name _____

At the Farmer's Market

You are bringing vegetables from your farm to the farmer's market in the city. The truck has room for 12 crates of vegetables.

1. Decide how to put the crates into the truck. Draw a picture to show how many crates you can put in 4 rows.

Workspace

2. How did you decide how many crates to put in 4 rows?

_____

3. Write the division fact that shows how many crates you put in 4 rows.

_____

4. What if your stand at the market can hold 8 crates? Draw a picture to show how many crates you could put in 2 rows.

Name _____

What is the shortest distance between two places?

There are many ways to travel from one place to another.

What to do

1. Work in 4 teams.

2. Use the diagram below to set up each team path. Mark each path with tape.

3. Have a volunteer make a tally for the winner of each race.

4. Repeat until all team members have raced.

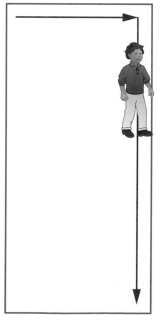

| Team 1 | Team 2 | Team 3 | Team 4 |

Fastest Path Tally Chart

Team 1	Team 2	Team 3	Team 4

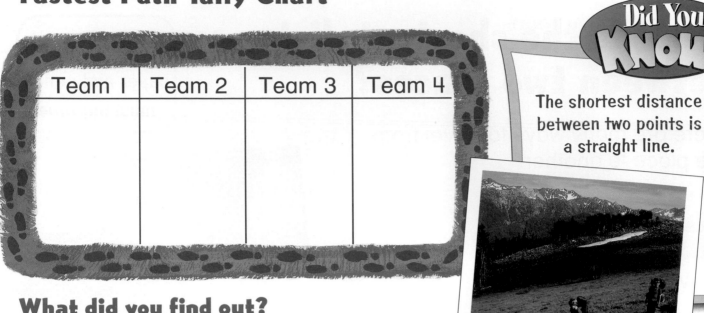

What did you find out?

1. Which team won the most races?

_____

2. Can you use multiplication to find how many children in all raced along the paths? Tell why or why not.

_____

3. How did the path affect which team won most often?

_____

4. What can you tell about the shortest distance between two places?

_____

 Want to do more?

Set up a path along a straight line and a path with curves. Race against a friend and see which one is shorter.

Check Your Progress B

Name _____

Find how many groups.

1. 12 boats

3 in each pond

_____ ponds

Find how many in each group.

2. 10 people

2 cars

_____ in each car

Make equal groups.
Write how many are in each group
and how many are left over.

3. 14 people

3 cars

_____ people in each car

_____ people left over

18 eggs

3 baskets

_____ eggs in each group

_____ eggs left over

Problem Solving

Draw a picture. Solve.

Workspace

4. 8 trucks each delivered 4 crates
of berries. How many crates
are there in all?

_____ crates

Name _____

Two Spins

- Label a spinner 2, 5, and 10.
  Label another spinner 0 – 9.

- You and a partner take turns.
  Choose yellow or red for your counters.

- Spin both spinners.

- Multiply the two numbers and place your counter
  on the product.

- The first player to cover a row, column, or diagonal wins.

You Will Need

2 spinners
2 color counters

Products

8	2	40	14	25
50	10	30	18	6
30	5	12	60	35
12	4	80	15	90
20	70	16	10	45

Name _____

Language and Math

Complete.
Use a word from the list.

Read these words.

Math Words

divide
factor
multiply
product
remainder

1. You _____ to make equal groups.

2. When you make equal groups and there is a leftover,

the leftover is called the _____.

3. In $3 \times 5 = 15$, 15 is the _____.

4. There are 10 trucks. Each truck carries 7 pigs.

_____ to find the total number of pigs.

Concepts and Skills

Find each product.

5. $2 \times 5 =$ _____ $5 \times 8 =$ _____ $5 \times 4 =$ _____

6.

$$\begin{array}{r} 5 \\ \times\, 3 \\ \hline \end{array} \qquad \begin{array}{r} 2 \\ \times\, 6 \\ \hline \end{array} \qquad \begin{array}{r} 5 \\ \times\, 7 \\ \hline \end{array} \qquad \begin{array}{r} 5 \\ \times\, 9 \\ \hline \end{array} \qquad \begin{array}{r} 10 \\ \times\, 4 \\ \hline \end{array} \qquad \begin{array}{r} 2 \\ \times\, 10 \\ \hline \end{array}$$

Chapter Review

Find how many groups.

7. 10 apples

   2 in each basket

   _____ baskets

8. 16 oranges

   4 on each tree

   _____ trees

Find how many in each group.

9. 12 pumpkins

   2 trucks

   _____ in each truck

10. 8 peaches

   4 baskets

   _____ in each basket

Make equal groups.
Write how many are in each group and how many are left over.

11. 11 into 4 equal groups: _____ in each group, _____ left over

12. 16 into 3 equal groups: _____ in each group, _____ left over

Problem Solving

Draw a picture.

13. 4 trucks each delivered 6 crates of berries. How many crates are there in all? _____

 Tell how a picture helps solve the problem.

Name _____

Find each product.

1. $3 \times 5 =$ ____ $5 \times 6 =$ ____ $5 \times 2 =$ ____

2. $2 \times 6 =$ ____ $2 \times 8 =$ ____ $3 \times 4 =$ ____

3.
$\begin{array}{r} 2 \\ \times\, 3 \\ \hline \end{array}$ $\begin{array}{r} 2 \\ \times\, 7 \\ \hline \end{array}$ $\begin{array}{r} 5 \\ \times\, 7 \\ \hline \end{array}$ $\begin{array}{r} 5 \\ \times\, 8 \\ \hline \end{array}$ $\begin{array}{r} 10 \\ \times\, 2 \\ \hline \end{array}$ $\begin{array}{r} 2 \\ \times\, 9 \\ \hline \end{array}$

Write how many in each group.

4. 10 boats

   5 harbors

   ____ in each harbor

5. 9 boats

   3 harbors

   ____ in each harbor

Make equal groups.
Write how many are in each group and how many are left over.

6. 15 into 4 equal groups: ____ in each group, ____ left over

7. 17 into 3 equal groups: ____ in each group, ____ left over

8. 14 into 7 equal groups: ____ in each group, ____ left over

16	25	50
30	8	12
22	10	45

Number of Horses

2	5	10
5	10	2
10	5	2

Number of Trucks

Toss a ▣ onto the green workmat.
Write this number in the purple box.

Toss a ▣ onto the truck workmat.
Write this number in the yellow box.

Make equal groups.
Write how many horses are in each truck and how many are left over.
Repeat.

horses	trucks	horses in each truck	leftovers

Name _____

Find each product.

1. $2 \times 2 =$ _____ $2 \times 8 =$ _____ $5 \times 4 =$ _____

2. $5 \times 7 =$ _____ $10 \times 3 =$ _____ $2 \times 10 =$ _____

3.
$$\begin{array}{cc} 3 \\ \times\,2 \\ \hline \end{array}$$
$$\begin{array}{cc} 5 \\ \times\,5 \\ \hline \end{array}$$
$$\begin{array}{cc} 2 \\ \times\,9 \\ \hline \end{array}$$
$$\begin{array}{cc} 10 \\ \times\,6 \\ \hline \end{array}$$
$$\begin{array}{cc} 10 \\ \times\,2 \\ \hline \end{array}$$
$$\begin{array}{cc} 5 \\ \times\,9 \\ \hline \end{array}$$

Find how many groups.

4. 12 airplanes

   3 on each ship

   _____ ships

Find how many in each group.

5. 12 people

   2 vans

   How many people are in each van? _____ in each van

Make equal groups.
Write how many are in each group and how many are left over.

6. 9 into groups of 4: _____ groups, _____ left over

Draw a picture. Solve.

7. 3 trucks each deliver 8
   crates of berries. How many
   crates are there in all?

   _____ crates

Workspace

You Will Need

2 cubes
counters

Number of bags   Number of Marbles

Toss a cube on each mat to find the number of bags and marbles.
Use the numbers to find how many marbles in all.
Write a multiplication sentence.
You can use counters if you need to.

1. _____ × _____ = _____   marbles in all

2. _____ × _____ = _____   marbles in all

3. _____ × _____ = _____   marbles in all

4. _____ × _____ = _____   marbles in all

Portfolio

You may want to put this page in your portfolio.

Name _____

Choose the correct answer.

Measurement and Geometry

1. Which shape has exactly 3 corners?

○ ⬠
○ ●
○ ■
○ ▲

2. What is the perimeter of the yellow rectangle?

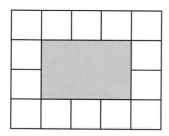

○ 3
○ 5
○ 6
○ 10

3. How many feet are there in a yard?

○ 1
○ 3
○ 10
○ 12

Statistics, Data Analysis, and Probability

4. What is the mode of these numbers?

3, 0, 0, 4, 10, 9, 12

○ 0
○ 10
○ 12
○ 0 to 12

5. Which is the most popular?

Favorite Way to Travel				

○ Car
○ Train
○ Bus
○ Airplane

6. If Mary takes a cube from the bag without looking, what cube is she most likely to pick?

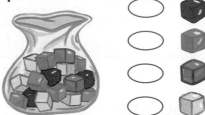

○ ▪
○ ▪
○ ▪
○ ▫

Mathematical Reasoning

7. The bus had 12 passengers. 5 got out. Which picture shows how many are left?

- ⬭ ●●●●●●●●●●●● ●●●●●
- ⬭ ✗●●●●●●●●●●●●
- ⬭ ✗✗✗✗✗●●●●●●●
- ⬭ ✗✗✗✗✗✗✗✗✗✗✗✗

8. Carl had $8. He spent $4. Which sentence shows how much money he has left?

- ⬭ $8 + $4 = $12
- ⬭ $8 − $4 = $4
- ⬭ $8 × $4 = $32
- ⬭ $4 + $8 = $12

9. Jim wants to divide bus tokens into 2 equal groups. Which is true?

- ⬭ There will never be leftovers.
- ⬭ There will always be leftovers.
- ⬭ There will be no leftovers if the number is even.
- ⬭ There will be no leftovers if the number is odd.

Number Sense

10. What is the value of the 5 in 450?

- ⬭ 5
- ⬭ 50
- ⬭ 55
- ⬭ 500

11. Add.

$$\begin{array}{r} 549 \\ + 275 \\ \hline \end{array}$$

- ⬭ 714
- ⬭ 814
- ⬭ 824
- ⬭ 5765

12. Subtract.

$$\begin{array}{r} 801 \\ - 583 \\ \hline \end{array}$$

- ⬭ 382
- ⬭ 302
- ⬭ 228
- ⬭ 218

Picture Glossary

add (+)

$$2 + 3 = 5$$

$$\begin{array}{r} 2 \\ +\ 3 \\ \hline 5 \end{array}$$

area

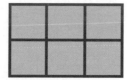

The **area** is 6 square units.

addend

$$\begin{array}{r} 31 \leftarrow \text{addend} \\ +\ 18 \leftarrow \text{addend} \\ \hline 49 \end{array}$$

array

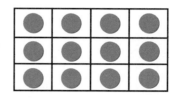

$$3 \times 4 = 12$$

addition sentence

$$3 + 8 = 11$$

bar graph

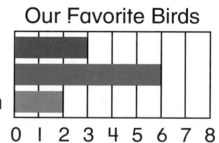

Our Favorite Birds

Robin
Blue jay
Goldfinch

0 1 2 3 4 5 6 7 8

after

47 48 49

48 is **after** 47

before

47 48 49

47 is **before** 48

angle

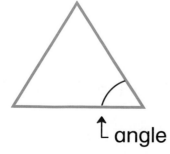

↑ angle

between

47 48 49

48 is **between** 47 and 49

Picture Glossary

calender

compare

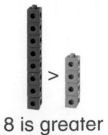

5 is less than 7

6 is equal to 6

8 is greater than 4

cent ¢

 or

1 ¢   1 cent

cone

centimeter

1 cm

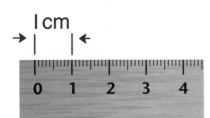

congruent

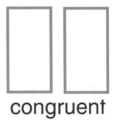

congruent

chart

Favorite Sports	
Soccer	7
Basketball	5
Football	4
Baseball	2

count back

3 - 1 = 2

circle

count on

3 + 1 = 4

Picture Glossary

counting pattern

2, 4, 6, 8, 10, 12, 14	Counting by 2's
3, 6, 9, 12, 15, 18, 21	Counting by 3's
5, 10, 15, 20, 25, 30, 35	Counting by 5's

decimal point

decimal point

cube

degree Celsius (˚C)

0°C

degrees ↗
Celsius (˚C)

cup

I Cup

degree Fahrenheit (˚F)

16°F

degrees ↗
Fahrenheit (˚F)

cylinder

difference

$$17 - 9 = 8$$

$$
\begin{array}{r}
17 \\
- 9 \\
\hline
8
\end{array}
$$

difference →

8 is the **difference** between 17 and 9

data

Information that is collected.

digit

3   5   4

↑   ↑   ↑

digits

Picture Glossary

dime or 10¢ 10 cents	**doubles** 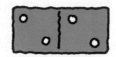 2 + 2 = 4
divide ÷  8 ÷ 4 = 2	**edge**  ↑ edge
division sentence 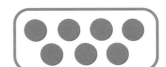 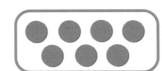 14 ÷ 2 = 7	**eighths** $\frac{1}{8}$ $\frac{1}{8}$ $\frac{1}{8}$ $\frac{1}{8}$ $\frac{1}{8}$ $\frac{1}{8}$ $\frac{1}{8}$ $\frac{1}{8}$ 8 **eighths** make a whole.
dollar  $1.00	**equal groups** 
dollar sign $	**estimate** 47 + 22 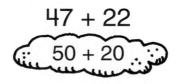 50 + 20 about 70 ← **estimate**

Picture Glossary

even

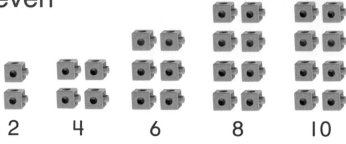

2   4   6   8   10

fair share

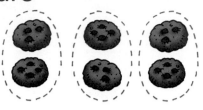

fair shares

expanded form

364

300 + 60 + 4

foot

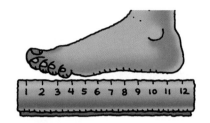

12 inches equal 1 **foot**.

face

face

fourths

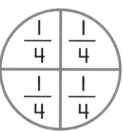

4 **fourths** makes 1 whole.

fact family

5 + 1 = 6   6 - 5 = 1

1 + 5 = 6   6 - 1 = 5

fraction

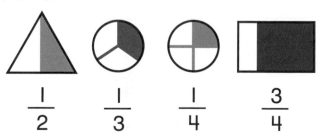

$\frac{1}{2}$ $\frac{1}{3}$ $\frac{1}{4}$ $\frac{3}{4}$

factor

2 x 3 = 6

factor

gram (g)

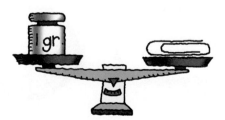

1 **gram** is about 1 paper clip.

Picture Glossary

half dollar $0.50  50¢ 50 cents	**hour hand**  hour hand →
half hour 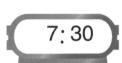 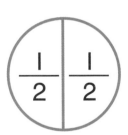 7:30 30 minutes	**hundreds** 2 3 4 ↑ 2 hundreds
halves  $\frac{1}{2}$ $\frac{1}{2}$ 2 **halves** makes 1 whole.	**inch** **(in.)** |← 1 in. →| 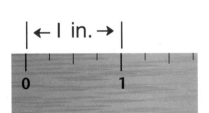 0 1
hexagon **Hexagon:** 6 sides	**is equal to** **(=)** 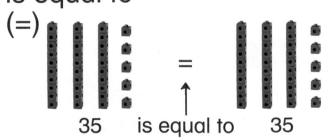 35 is equal to 35
hour   7:00 60 minutes	**is greater than** **(>)** 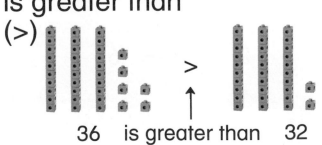 36 is greater than 32

Picture Glossary

is less than (<)

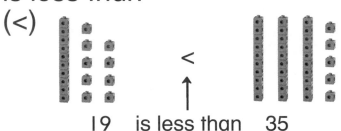

19   is less than   35

meter (m)

1 **meter** = 100 centimeters
1 **meter** is a little longer than a baseball bat.

kilogram (kg)

1 **kilogram** is about 4 oranges.

minute

60 seconds equal 1 **minute**.

length

Length is how long something is.

minute hand

minute hand

minute hand

line of symmetry

Symmetry: fold it and it matches

mode

4   7   10   36   7   2

most often

mode = most often = 7

liter (l)

month

This calendar shows the **month** of September.

Picture Glossary

multiplication sentence (X)

2 x 4 = 8

multiplication sentence

nickel

5¢   5 cents

number line

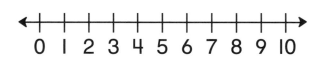

0 1 2 3 4 5 6 7 8 9 10

odd

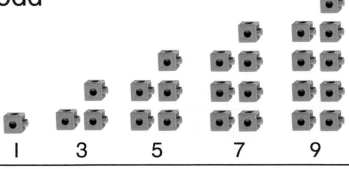

1   3   5   7   9

one eighth

$\frac{1}{8}$

One eighth is red.

one fourth

$\frac{1}{4}$

One fourth is red.

one half

$\frac{1}{2}$

One half is red.

one sixth

$\frac{1}{6}$

One sixth is red.

one third

$\frac{1}{3}$

One third is red.

one twelfth

$\frac{1}{12}$

One twelfth is red.

Picture Glossary

ones

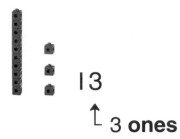

13

↑ 3 **ones**

penny ¢

1¢   1 cent

order

57, 58, 59, 60, 61

These numbers are in **order.**

pentagon

Pentagon: 5 sides

ordinal numbers

Jan is **seventh** in line.

perimeter

the distance around a shape

ounce (oz)

One CD weighs about 1 **ounce.**

pictograph

Games Won	
Bull Dogs	
Lions	

Each stands for 1 game.

parallelogram

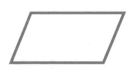

Parallelogram: 4 sides, 2 parallel, 2 same length

pint

2 cups = 1 **pint**

Picture Glossary

pound (lb)

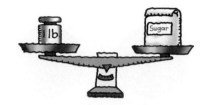

weighs about **1 pound**

quarter

or

25¢   25 cents

product

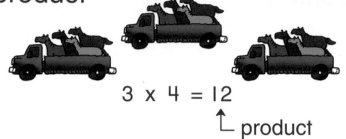

3 x 4 = 12
└ product

range

4   7   10   36   7   2
                    ↑   ↑
              greatest smallest

greatest minus smallest = **Range**
36 - 2 = 34

pyramid

rectangle

quadrilateral

quadrilateral: 4 sides

rectangular prism

quart

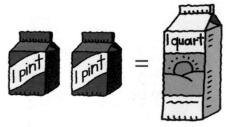

2 pints = **1 quart**

regroup

12 ones = 1 ten 2 ones

Picture Glossary

related facts

$$5 + 1 = 6$$

$$1 + 5 = 6$$

$$9 - 3 = 6$$

$$9 - 6 = 3$$

sphere

set

└→ set of numbers

square

side

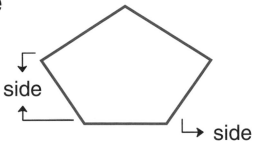

side

└→ side

subtract
(−)

$$47 - 28 = 19$$

sixths

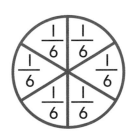

6 **sixths** makes 1 whole

sum

$$3 + 2 = 5$$ Add to find the **sum**.

$$3 + 2 = 5$$

└ 5 is the **sum** of 3 plus 2

skip count

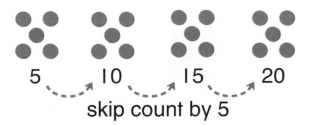

5   10   15   20

skip count by 5

survey

Favorite Sports	
Soccer	卌 II
Basketball	卌
Football	IIII
Baseball	II

This survey shows Favorite Sports.

Picture Glossary

symmetry

Symmetry: fold it and both halves match.

trapeziod

Trapezoid: 4 sides, 2 parallel, 2 not

tally mark

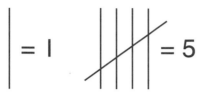

A **mark** is used to record data.

triangle

temperature

Temperature = 78°F

turnaround facts

$$\begin{array}{r} 4 \\ + 3 \\ \hline 7 \end{array} \qquad \begin{array}{r} 3 \\ + 4 \\ \hline 7 \end{array}$$

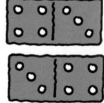

tens

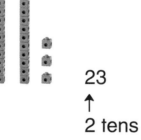

23

↑
2 tens

twelfths

12 **twelfths** makes 1 whole.

thirds

3 **thirds** makes 1 whole

vertex

Vertex

Picture Glossary

week

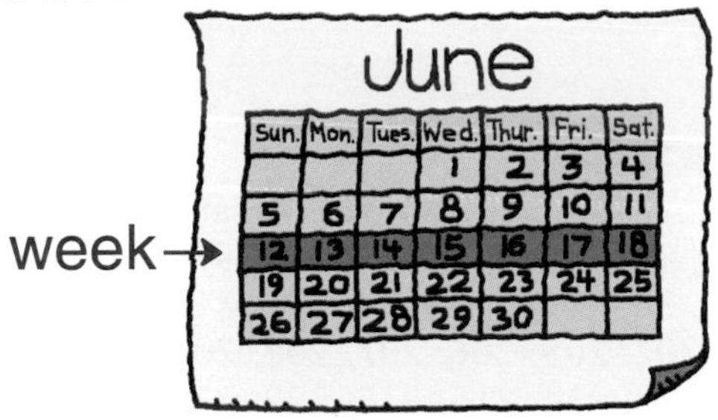

week →

yard (yd)

3 feet equal 1 **yard**.

year

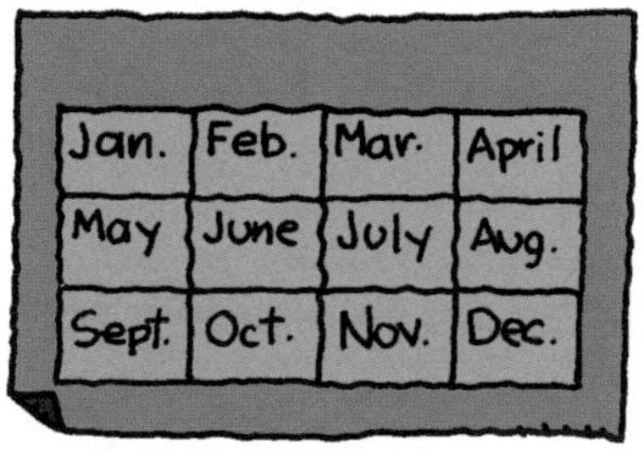

565

Credits